Lecture Notes in Earth Sciences

Edited by Somdev Bhattacharji, Gerald M. Friedman,
Horst J. Neugebauer and Adolf Seilacher

36

Dongkai Shangguan

Cellular Growth of Crystals

Springer-Verlag

Berlin Heidelberg New York London Paris
Tokyo Hong Kong Barcelona Budapest

Author
Dongkai Shangguan
University of Alabama, College of Engineering
F3101 Mineral Industries Building
Box 870202, Tuscaloosa, AL, 35487-0202, USA

ISBN 3-540-54485-2 Springer-Verlag Berlin Heidelberg New York
ISBN 0-387-54485-2 Springer-Verlag New York Berlin Heidelberg

This work is subject to copyright. All rights are reserved, whether the whole or part of the material is concerned, specifically the rights of translation, reprinting, re-use of illustrations, recitation, broadcasting, reproduction on microfilms or in other ways, and storage in data banks. Duplication of this publication or parts thereof is only permitted under the provisions of the German Copyright Law of September 9, 1965, in its current version, and a copyright fee must always be paid. Violations fall under the prosecution act of the German Copyright Law.

© Springer-Verlag Berlin Heidelberg 1991
Printed in Germany

Typesetting: Camera ready by author
Printing and binding: Druckhaus Beltz, Hemsbach/Bergstr.
32/3140-543210 – Printed on acid-free paper

To
My Parents
and
My Wife Guilian

Preface

This book is based on the author's D. Phil. Thesis which was submitted to the University of Oxford in October 1988. The work described in the book was carried out by the author in the Department of Metallurgy and Science of Materials at the University of Oxford, between October 1985 and September 1988, under the supervision of Dr. John D. Hunt. The research described here is original; where the work of others has been drawn upon, it has been acknowledged in the text and a list of references is included at the end of the book.

I would like to thank all those who have helped and encouraged me during the course of this work, and in particular, Dr. John D. Hunt for his patient advice, constant encouragement and inspiring supervision; Drs. Brian Cantor and Donald T.J. Hurle, my D. Phil. Examiners, for their constructive criticisms; Professor Sir Peter Hirsch, F.R.S. and Professor J.W. Christian, F.R.S., for the provision of laboratory facilities; The Oxford University Computing Service for the provision of computing facilities; The State Education Commission of China and the British Council for financial support (1985-1988); University of Oxford for the award of a Scholarship for Overseas Students (1985-1988); The Principal and Fellows of St. Edmund Hall for the award of a Brockhues Graduate Award (1986-1988); Mr. John Short of the Student's Workshop for his advice on practical matters; Mr. A.H. McKnight for photographic work; and the technical and secretarial staff of the Department for their willing assistance at all times. I would also like to thank my many friends who have offered me friendship and moral support over the years at Oxford, especially Drs. Stephen Flood, Melinda Bagshaw, Giles Rodway, Y.K. Park, and Mehmet Gunduz, as well as Hailin Jiang, Miss Alison Kent, and Firuz Guven. Finally, I would like to thank my parents and my wife Guilian, to whom this book is dedicated, for their constant support, encouragement, patience and interest in my work.

Dongkai Shangguan
Tuscaloosa, Alabama
April 1991

ABSTRACT

Cellular growth is an important crystal growth process and offers an interesting example of natural pattern formation. The present work has been undertaken to study cellular growth, especially its pattern formation, both experimentally and numerically. *In situ* observations of faceted cellular growth clearly revealed cellular interactions in the array of cells. Cell tip splitting and loss of cells were observed to be the two main mechanisms for the adjustment of cell spacings during growth. For the first time, the true time-dependent faceted cellular growth has been modelled properly. The time evolution of faceted cellular growth has demonstrated the dynamical features of cellular growth processes. It was shown that the pattern formation was determined by cellular interactions in the array, either transient or persistent depending on the growth condition. The cellular structures were irregular when persistent interactions occurred, whereas relatively regular structures could be formed once the transient interactions had stopped. As a result of cellular interactions, a finite range of stable cell spacings was found under a given growth condition. Numerical experiments were carried out for $k > 1$ and $k < 1$ (where k is the solute partition coefficient), under a number of different growth conditions. It was found that these two cases were not symmetric as far as solute distribution is concerned; however the pattern formation behaviours were similar. For $k > 1$ shallow cells were retained, while for $k < 1$, the formation of liquid grooves along the cell boundary depended on the growth condition. The solute effect plays an important role in the cellular interactions in the array. The results were compared with experimental observations in thin film silicon single crystals. It is felt that a general behaviour of pattern formation is found and should be expected for other processes such as non-faceted cellular or eutectic growth. In addition, the solute flow in steady state cellular array growth was studied using the point source technique. Preliminary work was carried out to measure steady state non-faceted cell shapes. Heat flow in zone melting was studied numerically.

Index to Sources

Figure	Source	Page
2.1	K.A. Jackson, <u>Liquid Metals and Solidification</u>, American Society for Metals, 1958. *Reproduced with permission.*	5
2.2 & 2.8	D.P. Woodruff, <u>The Solid-Liquid Interface</u>, Cambridge University Press, 1973. *Reproduced with permission.*	5,10
2.3 & 2.9	J.W. Christian, <u>The Theory of Transformations in Metals and Alloys</u>, Pergamon Press, 1965. *Reproduced with permission.*	7,12
2.4 & 2.7	M.C. Flemings, <u>Solidification Processing</u>, McGraw-Hill, 1974. *Reproduced with permission.*	7,9
2.10	W.K. Burton, N. Cabrera & F.C. Frank, <u>Philosophical Transactions</u>, vol. A243, The Royal Society, 1950. *Reproduced with permission.*	12
2.11	J.W. Cahn, W.B. Hillig & G.W. Sears, <u>Acta Metallurgica</u>, vol. 12, Pergamon Press, 1964. *Reproduced with permission.*	13
2.12	C. Herring, <u>Structure and Properties of Solid Surfaces</u> (Eds. R. Gomer & C. Smith), The University of Chicago Press, 1953. *Reproduced with permission.*	13
2.13	L. Pfeiffer, S. Paines, G.H. Gilmer, W. Saarloos & K.W. West, <u>Physical Review Letters</u>, vol. 54(17), American Physical Society, 1985. *Reproduced with permission.*	20
4.10	W.G. Pfann, <u>Zone Melting</u>, 2nd Edition, John Wiley & Sons, 1966. *Reproduced with permission.*	38,39
4.10(4)	<u>Journal of Metals</u>, vol. 6(9), The Minerals, Metals, and Materials Society, 1954. *Reproduced with permission.*	39
9.33	P.M. Zavracky, D.P. Vu, L. Allen, W. Henderson, H. Guckel, J.J. Sniegowski, T.P. Ford & J.C.C. Fan, <u>Silicon-On-Insulator and Buried Metals in Semiconductors</u> (Eds. J.C. Sturm, C.K. Chen, L. Pfeiffer & P.L.F. Hemment), Materials Research Society Symposium Proceedings, vol. 107, Materials Research Society, 1988. *Reproduced with permission.*	198

CONTENTS

CHAPTER I. Introduction   1

CHAPTER II. Literature Review

2.1	Interface Structure and Kinetics	2
2.1.1	Atomic process at the solid / liquid interface	2
2.1.2	Interface structure	2
2.1.2.1	Jackson's theory	2
2.1.2.2	Cahn's theory	4
2.1.2.3	Faceted growth and non-faceted growth	6
2.1.3	Interface kinetics	6
2.1.3.1	Continuous growth	6
2.1.3.2	Lateral growth	8
	(1) Growth by two-dimensional nucleation	8
	(2) Growth by screw dislocations	8
2.1.3.3	A unified theory	11
2.2	Growth of the Interface	14
2.2.1	Interface condition	14
2.2.2	Heat and solute transport	15
2.2.3	Planar front stability	16
2.2.3.1	Constitutional undercooling	16
2.2.3.2	Mullins-Sekerka theory	16
2.2.4	The growth condition	17
2.2.4.1	Isolated dendritic growth	17
2.2.4.2	Array growth	17
2.2.5	Steady state non-faceted cellular array growth	17
2.2.6	Stability treatment	17
2.2.6.1	Extremum growth hypothesis	18
2.2.6.2	Marginal stability hypothesis	18
2.2.7	Experimental work	18
2.2.8	Dendritic growth	19
2.2.9	Faceted cellular array growth	19
2.2.10	Pattern formation	21
2.3	Objective of the Present Work	21

CHAPTER III. Introduction to Numerical Modelling

3.1	Numerical Solution of Differential Equations	22
3.2	Finite Difference and Finite Element Methods	22
3.3	The Formulation of Discretisation Equations	22
3.4	Grids	23
3.5	Interpolation (Extrapolation)	23
3.6	The Steady State Problem and Time-Dependent Problem	23

3.7	Convergency, Consistency and Stability	26
3.7.1	Convergency	26
3.7.2	Consistency	26
3.7.3	Stability	26
3.8	Other Numerical Methods	27

CHAPTER IV. Experimental Techniques and Apparatus

4.1	Introduction	28
4.2	The Temperature Gradient Stage	28
4.2.1	General considerations	28
4.2.2	The design and construction	28
4.2.2.1	The main body	28
4.2.2.2	Motion of the specimen	28
4.2.2.3	Heating	31
4.2.2.4	The stage	31
4.2.3	Calibrations	31
4.2.3.1	The moving rate of the specimen	31
4.2.3.2	The temperature gradient	31
4.2.4	Operation and performance	33
4.3	Material Selection and Purification	33
4.3.1	Material selection	33
4.3.2	Material purification	33
4.3.2.1	Vacuum distillation	33
4.3.2.2	Zone refining	37
	(1) Introduction	37
	(2) Principles of zone refining	37
	(3) Construction of the zone refiner	40
	(4) Preparation of the sample	43
	(5) Operation	44
	(6) Results of zone refining	44
4.4	Preparation of the Specimen	44
4.4.1	The glass cell	44
4.4.2	The specimen	46
4.5	Optical Considerations	46
4.6	The Experimental Procedure	46
4.7	Summary	48

CHAPTER V. Experimental Results

5.1	Introduction	49
5.2	Faceted Cellular Growth	49
5.2.1	Introduction	49
5.2.2	Results	49
5.2.2.1	Development of faceted cellular arrays	49

5.2.2.2	Solute redistribution during growth	52
5.2.2.3	Morphology of the interface	57
5.2.2.4	Growth competition between facets	57
5.2.2.5	Cell interactions: formation of new cells	57
5.2.2.6	Cell interactions: loss of cells	57
5.2.2.7	Cell spacings	67
5.2.3	Discussions and summary	67
5.3	Preliminary Work with Non-Faceted Cellular Growth	70
5.3.1	Introduction	70
5.3.2	Results and discussion	70
5.3.3	Summary	74

CHAPTER VI. Numerical Study of Heat Flow in Zone Melting

6.1	Introduction	76
6.2	The Problem	76
6.3	The Mathematical Formulation	76
6.3.1	The heat flow equation	76
6.3.2	The discretization equations	78
6.3.3	Boundary conditions	80
6.3.4	The enthalpy method	83
6.3.5	Formulation of an implicit scheme	85
6.3.6	Newton-Raphson method	86
6.4	The Computational Procedure	88
6.5	Results and Discussion	88
6.6	Summary	90

CHAPTER VII. Theoretical Study of Steady State Cellular Array Growth

7.1	Introduction	92
7.2	The Problem	92
7.3	The Technique	94
7.4	The Mathematical Formulation	95
7.4.1	The general scheme	95
7.4.2	Normalization of variables	95
7.4.3	Construction of the Green's function for the problem stated by Eqs. (7.14)-(7.16)	96
7.4.4	Solution of $G(X^*, Z^*, X'^*, Z'^*)$	96
7.4.4.1	Application of the finite Fourier cosine transform	96
7.4.4.2	Application of the complex Fourier transform	97
7.4.4.3	Solution of $G(X^*, Z^*, X'^*, Z'^*)$	98
7.4.5	Introduction of the function $F(X^*, Z^*)$	99
7.4.6	Construction of the Green's function $G(X^*, Z^*, X'^*, Z'^*)$ for $F(X^*, Z^*)$	100
7.4.7	Application of the Green's theorem	100
7.4.8	Reversion to $C^*(X^*, Z^*)$	102

7.5	Numerical Evaluation of Analytical Expressions	105
7.5.1	Evaluation of the Green's function and its derivatives	105
7.5.2	Solution of the solute field	108
7.5.2.1	The integral equation	108
7.5.2.2	Discretization	112
7.5.2.3	Linearization of the equation	112
7.5.2.4	Solution for the interface concentration	113
7.5.3	The computational effort	113
7.5.3.1	Numerical integration	113
7.5.3.2	Solution of linear equations	113
7.6	Results	113
7.7	Discussions	120
7.8	Summary	121

CHAPTER VIII. Numerical Dynamical Study of Faceted Cellular Array Growth (Part I)

8.1	Introduction	123
8.2	The Physical Model	123
8.3	The General Scheme of the Numerical Work	125
8.4	Model I	126
8.4.1	Description of the model	126
8.4.2	Treatment of tip splitting and loss of cells	128
8.4.2.1	Tip splitting	128
8.4.2.2	Loss of cells	128
8.4.3	Results	128
8.4.4	An analytical model of the steady state cell spacing	128
8.4.5	Discussions and conclusions	134
8.5	Model II	134
8.5.1	Description of the model	134
8.5.2	Results and discussions	135
8.6	Model III	135
8.7	Summary	139

CHAPTER IX. Numerical Dynamical Study of Faceted Cellular Array Growth (Part II)

9.1	Introduction	140
9.2	Model IV	140
9.2.1	Description of the model	140
9.2.2	The modelling procedure	141
9.2.3	The initial condition	141
9.2.4	The grid	143
9.2.5	Growth of the cells	143
9.2.6	Grid for the new interface	143

9.2.7	Solution of solute diffusion during growth	146
9.2.8	Formation and treatment of liquid grooves	153
9.2.8.1	Position of the liquid groove	153
9.2.8.2	Solute flow along the groove	153
9.2.9	Treatment of tip splitting and loss of cells	153
9.2.9.1	Loss of cells	153
9.2.9.2	Tip splitting	154
9.2.10	Calculation of the solute profile at new grid points	154
9.2.11	Computational considerations	154
9.2.11.1	Choice of the time step	154
9.2.11.2	Monitoring the calculation	156
9.2.11.3	The computational efforts	156
9.3	Results and Discussions	156
9.3.1	Faceted cellular array growth	156
9.3.2	Cellular interactions	186
9.3.3	The stable cell spacing range	186
9.3.4	Solute redistribution	189
9.3.5	Deep cells and shallow cells	189
9.3.6	The effect of solute on cellular interactions	189
9.3.7	Comparison with experimental observations	194
9.4	General Discussions	194
9.5	Pattern Formation in Perspective	199
9.6	Conclusion	199

CHAPTER X. Summary of Conclusions and Suggestions for Future Work

10.1	Conclusions	200
10.2	Suggestions for Future Work	200

REFERENCES	202
SUBJECT INDEX	209

Chapter I

INTRODUCTION

When a crystal undergoes a liquid/solid transformation, *i.e.* solidification, a planar interface is first formed between the solid phase and the liquid phase. If the planar interface is unstable against natural perturbations, as is the case in most practical situations, cells will be formed. Further away from the planar front stability condition, dendrites are formed. The solidification structure of a casting provides the basis for any further treatment, and primarily determines the physical, mechanical, and chemical properties of the resultant products. It is therefore very important to study the solidification process in order to be able to control the solidification microstructure and to achieve the desired material properties. The study of cellular growth is important, because the cellular/dendritic structure is the final structure of most castings, and the study of cellular growth provides the basis for the study of dendritic growth. Cellular growth also offers an interesting example of natural pattern formation.

There has been considerable amount of work on cellular growth over the last few decades. The steady state non-faceted cellular growth problem has been addressed satisfactorily; however questions still remain to be answered on the natural pattern formation of cellular growth. More recently faceted cellular growth has aroused great interest because of its importance to the understanding of the crystal growth of several important semiconductor materials (Chapter 2).

This work has been undertaken to study cellular growth, especially its pattern formation. The direct observation of faceted cellular growth clearly revealed cellular interactions in the array of cells (Chapters 4&5). Cell tip splitting and loss of cells were observed to be the two main mechanisms for the adjustment of cell spacings during growth. For the first time, the true time-dependent faceted cellular growth has been modelled properly; the time evolution of faceted cellular growth has demonstrated the dynamical features of cellular growth processes (Chapters 8&9). It was shown that the pattern formation was determined by cellular interactions in the array, either transient or persistent depending on the growth condition. As a result, a finite range of stable cell spacings was found under a given growth condition. The cellular structures were irregular when persistent interactions occurred, whereas relatively regular structures could be formed once the transient interactions had stopped. Numerical experiments were carried out for $k > 1$ and $k < 1$ (where k is the solute partition coefficient), under a number of different growth conditions. It was found that these two cases were not symmetric as far as solute distribution is concerned; however the pattern formation behaviours were similar. The solute effect plays an important role in the cellular interactions in the array. The results were compared with experimental observations in thin film silicon single crystals. It is felt that a general behaviour of pattern formation is found and should be expected for other processes such as non-faceted cellular or eutectic growth.

In addition, preliminary work was carried out to measure the steady state non-faceted cell shapes (Chapter 5). Heat flow in zone melting was studied numerically (Chapter 6). Solute flow in steady state cellular array growth was studied using the point source technique (Chapter 7). An introduction to numerical modelling was given in Chapter 3.

Chapter II

LITERATURE REVIEW

2.1 Interface Structure and Kinetics

2.1.1 Atomic process at the solid/liquid interface

During growth, atoms (molecules, or clusters) move between the solid/liquid interface and the liquid phase. As they move towards and strike against the interface, they may stick to the interface or leave the interface and move back into the liquid. Solidification proceeds by the addition of atoms to the solid/liquid interface. The ease with which atoms can attach themselves to a growing solid interface depends on the atomic structure of the interface. Different interface structures lead to different growth kinetics and different growth morphologies.

2.1.2 Interface structure

2.1.2.1 Jackson's theory

The analysis of Jackson [1] is made on thermodynamical grounds. In this theory, the equilibrium interface structure is considered which is in equilibrium with both the crystal and the melt.

Consider an interface which is initially atomically smooth. Atoms are added to the interface in a single layer until a complete monolayer has been formed; the surface is then once again atomically smooth, but has advanced by one atom spacing. Now consider the change in free energy associated with adding a certain number of atoms randomly. At constant pressure the excess free energy may be written as:

$$\Delta G = -\Delta E_0 - \Delta E_1 + T\Delta S_0 - T\Delta S_1 - P\Delta V \qquad (2.1)$$

where ΔE_0, ΔE_1, ΔS_0, ΔS_1, and ΔV are all defined as positive, and ΔE_0 is the change in internal energy associated with the atoms being attached to the surface, ΔE_1 is the change in internal energy associated with the atoms on the surface due to the presence of other adatoms on the monolayer, ΔS_0 is the change in entropy associated with the adatoms passing to the solid phase from the liquid phase, ΔS_1 is the configurational entropy associated with the different possible sitings of the adatoms on the surface, and ΔV is the change in volume due to the atoms being associated with the new phase, which will subsequently be approximated to zero for the solid-liquid transition.

Now consider one adatom which has η_0 nearest neighbours in the solid which were present before any atoms are added, and a maximum possible number of adatom nearest neighbours η_1. The symmetry of the crystal structure is such that, if growth were to continue until the atom considered became situated within the bulk of the solid, it would gain a further

η_0 nearest neighbours in the process. Thus if in the bulk solid an atom has ν nearest neighbours, then

$$\nu = 2\eta_0 + \eta_1 \tag{2.2}$$

An atom transferred from bulk liquid to an isolated site on a singular surface will suffer a change in internal energy, $L_0(\eta_0/\nu)$, where L_0 is the change in internal energy associated with the transfer of one atom from bulk liquid to bulk solid (which, at the equilibrium temperature for the phase change, T_E, is equivalent to the heat of fusion per atom). The η_0 nearest neighbour atoms in the solid will each suffer a similar change in internal energy, L_0/ν. Therefore for N_A adatoms

$$\Delta E_0 = 2 L_0 \left(\frac{\eta_0}{\nu}\right) N_A \tag{2.3}$$

If N is the number of atoms in a complete monolayer on the surface considered, any adatom site will on average have N_A/N nearest neighbour adatom sites filled. Thus

$$\Delta E_1 = L_0 \left(\frac{\eta_1}{\nu}\right)\left(\frac{N_A}{N}\right) N_A \tag{2.4}$$

In this case no factor of 2 appears because the bonds considered here are only between adatoms and do not involve the substrate.

Also we have

$$\Delta S_0 = \left(\frac{L_0}{T_E}\right) N_A \tag{2.5}$$

The configurational entropy, ΔS_1, is given by $(k \ln W)$, where k is the Boltzmann constant, and W is the number of ways of arranging the N_A atoms in the N sites, which is given by

$$W = \frac{N!}{N_A! (N - N_A)!} \tag{2.6}$$

Approximately we get

$$\Delta S_1 = k N \ln\left(\frac{N}{N - N_A}\right) + k N_A \ln\left(\frac{N - N_A}{N}\right) \tag{2.7}$$

Assuming that

$$T = T_E \tag{2.8}$$

$$\alpha = \left(\frac{L_0}{k\,T_E}\right)\left(\frac{\eta_1}{v}\right) \tag{2.9}$$

combining the above equations gives

$$\frac{\Delta G}{N\,k\,T_E} = \frac{\alpha\,N_A\,(N - N_A)}{N^2} - \ln\!\left(\frac{N}{N - N_A}\right) - \frac{N_A}{N}\ln\!\left(\frac{N - N_A}{N_A}\right) \tag{2.10}$$

Eq. (2.10) may be displayed graphically as shown in Fig. 2.1 for different values of the parameter α. It can be seen that the curve has a minimum at $N_A/N = 0.5$ for all values of α less than 2, indicating that a rough interface, *i.e.* an interface with half the sites filled, should be the equilibrium form; whereas for all higher values of α, it has 2 minima, one at a very small value of N_A/N and the other for a value close to 1, indicating that a smooth interface should be the equilibrium form.

The parameter α that controls the structure of the equilibrium interface consists of two parts: $L_0/(kT_E)$, which depends on the material, and η_1/v, which depends on the crystal structure and the surface under consideration, being highest for the most closely packed planes.

Generally the success of the predictions of Jackson's analysis is very good [2,3]. Metals, the main low α factor materials, grow with non-singular interfaces, and most non-metals and compounds grow with a faceted interface from their melt. Most organic compounds have large entropies of melting and grow in a faceted manner. There is however a small group of organic compounds which have a very low entropy of melting and solidify in the same way that metals do. These offer convenient transparent analogous systems for the study of metallic solidification [4].

2.1.2.2 Cahn's theory

Although practically Jackson's analysis makes good predictions in most cases, it cannot explain the transition of some borderline materials from faceting to non-faceting as the interface undercooling is increased, because this analysis is essentially an equilibrium one. A different approach is that of Cahn [5] who concludes that it is not valid to discuss a surface as being singular or non-singular, or to discuss whether growth is by the lateral propagation of steps or the forward motion of the interface, without taking into account the effect of the driving force on the nature of the interface.

The basic premises of this theory are, firstly, at sufficiently low growth rates the interface will always attempt to be in equilibrium, and secondly, the free energy of an interface is a periodic function of its mean position relative to the lattice periodicity of the solid phase. The free energy of the interface as a function of position can be drawn schematically as in Fig. 2.2. If the applied driving force for growth is less than the free energy change between the maxima (Fig. 2.2(1)), then the interface cannot pass through the corresponding intermediate positions. However, it can advance by a lateral mechanism in which the mean position of the interface over each terrace only differs from the next by an integral number of monolayer spacings (*i.e.*, these

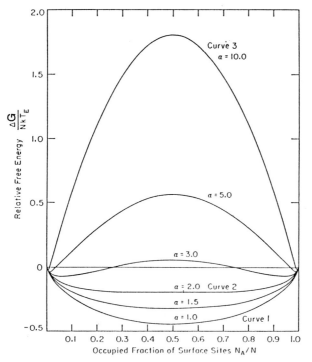

Fig. 2.1 Curves of excess free energy versus monolayer occupation for various values of the parameter α (from Ref.1).

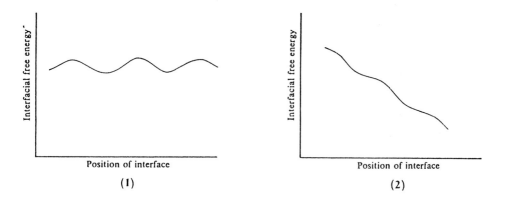

Fig. 2.2 Schematic graph of the free energy of an interface as a function of its mean position normal to the interface (after Ref. 3).

(1) At low driving force.

(2) At high driving force.

are all minimum free energy configurations apart from the small ledge energy). On the other hand, if the driving force is sufficiently high, the situation becomes more like Fig. 2.2(2); there is no longer a potential barrier to continuous growth.

The essence of this theory is that there is a continuous range of possibilities between completely smooth and completely rough, the criteria being both the material and the driving force. For any interface, there is a driving force below which growth is by lateral propagation of steps, and above which normal propagation occurs. The value of the critical driving force depends on the diffuseness of the interface, being very large for a perfectly smooth interface, and decreasing as the diffuseness increases.

This theory however does not give a method of calculating the 'diffuseness parameter'. It has also been criticized for the fact that it is based on a second-order phase transformation and does not therefore apply to solidification and melting which are first-order transformations [6].

2.1.2.3 Faceted growth and non-faceted growth

Different interfaces grow in different ways resulting in different interface morphologies. In the ideally diffuse interface, most lattice sites are favorable for atomic deposition; and so as atoms strike the interface it moves forward more or less uniformly and growth is said to be continuous, resulting in the non-faceted morphology of the interface. If, on the other hand, the interface is atomically flat, then forward growth occurs preferentially at steps which sweep laterally across the interface, and the growth is then said to proceed by lateral growth, resulting in the faceted morphology of the interface.

For faceted growth, because high index faces grow more rapidly than low index faces, high index faces will eventually disappear leaving the low index faces behind [7] (Fig. 2.3). For example, in silicon, germanium and other f.c.c systems, the facets developed are {111} planes.

2.1.3 Interface kinetics

As we can expect, different interfaces grow by different mechanisms and therefore at different rates. For example, we can expect that ideally diffuse interfaces should grow more easily than flat interfaces. The interface kinetics describe the relationship between the growth rate of the interface, R, and the driving force, *i.e.* the kinetic undercooling ΔT_k.

2.1.3.1 Continuous growth

For an ideally diffuse interface, as atoms strike the interface they may stay any sites at the interface, and the interface moves continuously. The growth rate of the interface is simply the difference between the jump rates of atoms to and from the interface. This is shown to be [8]

$$R = \beta \frac{D_L}{D_{LM}} \Delta T_k \qquad (2.11)$$

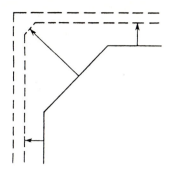

Fig. 2.3 Sketch showing how rapidly growing crystal faces disappear from the crystal form (after Ref. 7).

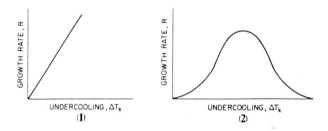

Fig. 2.4 Characteristic growth rate curves for (1) nonviscous liquids, such as metal, and (2) oxide or organic polymers (after Ref. 8).

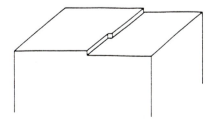

Fig. 2.5 Atomically flat interface with a step.

where β is a constant, D_L is the liquid diffusion coefficient, and D_{LM} is the liquid diffusion coefficient at the melting temperature.

If D_L does not change much with undercooling from D_{LM}, as in metals for example, the relationship between R and ΔT_k is linear (Fig. 2.4(1)). When D_L varies strongly with temperature, as in oxide glasses and polymers, R increases to a maximum at some undercooling and then decreases at higher undercoolings (Fig. 2.4(2)).

2.1.3.2 Lateral growth

Fig. 2.5 shows a flat interface with a step. Growth of the flat interface takes place by such steps sweeping across it. The source of steps however can be different leading to different growth mechanisms.

(1) Growth by two-dimensional nucleation [8]

One source of steps on low index faces is two-dimensional nucleation (Fig. 2.6). The problem of nucleus formation is closely analogous to the classical three-dimensional nucleation problem. As the growth rate of each nucleus which forms is sufficiently rapid relative to the crystal size, we can assume that one nucleus spreads to form a full atomic plane before the next nucleus forms. Then the growth rate of the interface becomes

$$R = A \, I_{2d} \, a \tag{2.12}$$

where A is the surface area of the growing face, I_{2d} is the rate of formation of new nuclei, and a is the height of the step, *i.e.*, the atomic spacing. Substituting the nucleation rate into Eq. 2.12, the growth rate can be written as

$$R = \beta \frac{D_L}{D_{LM}} \exp\left(-\frac{\beta'}{\Delta T_k}\right) \tag{2.13}$$

where β and β' are constants. This is schematically shown in Fig. 2.7. Typically the critical undercooling ΔT_k required to achieve an observable growth rate is much larger than is observed in most practical cases.

At higher undercoolings, the density of two-dimensional nuclei will become so high that many nuclei form for each plane to grow, and the growth rate will approach the continuous growth law as the upper limit.

(2) Growth by screw dislocations

Frank [9] first pointed out that a screw dislocation emerging at the interface provides a step in the interface (Fig. 2.8) and obviates the need for two-dimensional nucleation permitting growth at much lower undercoolings than that predicted by the two-dimensional nucleation

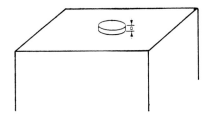

Fig. 2.6 A two-dimensional nucleus on a flat interface.

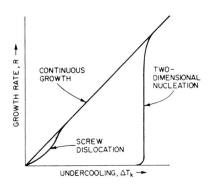

Fig. 2.7 Growth rate versus interface undercooling according to the three classical laws (after Ref. 8).

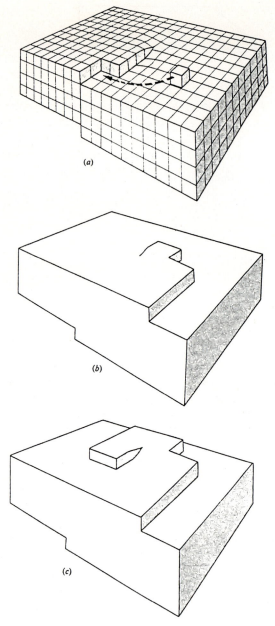

Fig. 2.8 Schematic drawing showing the point of emergence of a screw dislocation at the surface of a cubic solid, and of the development of a growth spiral when growth occurs on this surface (after Ref. 3).

theory. The step is self-perpetuating; no matter how many layers of atoms are deposited on the face, the step will persist.

For a single dislocation, the crystal grows up a spiral staircase by continuously depositing atoms at the exposed step. The step therefore rotates continuously about the point where the dislocation emerges. Since one end of the step is fixed, the step rapidly winds itself up (Fig. 2.9). At the centre of the spiral, it reaches a minimum radius of curvature which is the same as the critical radius for two-dimensional nucleation; at this curvature, the step edge is just in equilibrium with the surrounding melt and neither advances nor retreats. Further out along the spiral, the curvature is less and the step advances at a greater rate. At steady state, the wound-up spiral would appear to be rotating at a constant angular velocity.

Assuming the spiral to be archimedean, which has been shown [10] to give very nearly the correct final result, the growth rate is given [8] by

$$R = \beta \frac{D_L}{D_{LM}} (\Delta T_k)^2 \qquad (2.14)$$

where β is a constant. Thus we expect the square dependence of the growth rate on undercooling (Fig. 2.7). Growth approaches the continuous growth law as the upper limit.

The treatment leading to Eq. 2.14 has dealt with only a single dislocation. However real crystals may have a dislocation density as high as 10^8 lines/cm^2. Another complication is that the dislocations emerging at the interface may be with different undercoolings, *e.g.*, when there is a temperature gradient along the interface. Careful analysis of the interactions between the growth spirals [10] has shown that the growth rate of the interface is just that which would result if the kinetic undercooling over the entire interface were equivalent to that of the maximum, *i.e.*,

$$R \propto (\Delta T_k)^2_{max} \qquad (2.15)$$

where $(\Delta T_k)_{max}$ is the maximum kinetic undercooling along the interface.

Fig. 2.10 shows [10] the simplest case of multiple dislocations, where growth is proceeding from two dislocations of opposite sign spaced apart not less than 2r* (r* is the minimum radius); the growth rate however remains unchanged.

2.1.3.3 A unified theory

Cahn's theory [5] predicts that there is a continuous range of possibilities of interface structure between completely smooth (faceted) and completely rough (non-faceted), the criteria being the material and the driving force. This has been confirmed by more recent work on molecular dynamics simulation of interface structure [11-14]. Accordingly, the growth kinetics is expected to vary, continuously, with the variation of interface structure, as predicted by Cahn *et al.* [15] (Fig. 2.11).

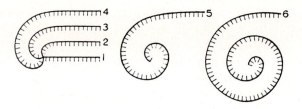

Fig. 2.9 Development of spiral structure (after Ref.7).

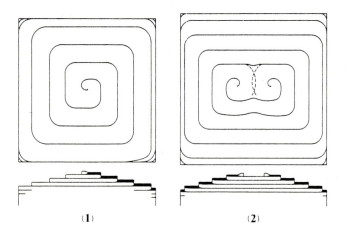

Fig. 2.10 Growth pyramid due to (1) a single dislocation and (2) a pair of dislocations (after Ref. 10).

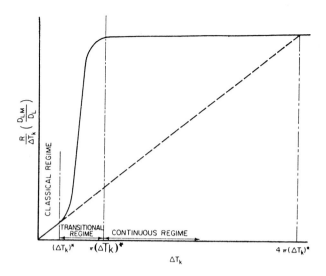

Fig. 2.11 Predicted growth rate curves for surface with an emergent dislocation. Ordinate is interface velocity divided by undercooling and corrected for temperature dependence of the diffusion coefficient (after Ref. 15).

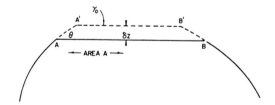

Fig. 2.12 Definition of the terms used in Eq. (2. 18): γ_0 represents surface tension of facet A'B' or AB; γ_1 represents surface tension of surrounding facets such as AA' or BB'; A represents area of facet AB; and θ is dihedral angle between facet AB and the bounding surface (after Ref. 22).

A way of testing the theory has been to measure the interface growth rate at different kinetic undercoolings and compare with the predictions. There has been considerable amount of experimental work in this area; for example, the results obtained by Abbaschian *et al.* [16] and Peteves *et al.* [17-19] have shown good agreement with theoretical predictions.

2.2 Growth of the Interface

Growth of the interface is determined by heat and solute transport, and the interface condition. Cellular and dendritic growth occurs following the breakdown of the planar solid/liquid interface. Stability considerations deal with the pattern formation of cellular/dendritic growth.

2.2.1 Interface condition

Under normal solidification conditions, the growth rate of the interface is much less than the rate of transportation of atoms, and local equilibrium is assumed to apply at the interface. The interface temperature T_I is related to the interface liquid concentration C_{LI}, the surface curvature, and the departure of the interface from local equilibrium, through the interface undercooling equation [20]:

$$\Delta T_I = T_0 - T_I = \Delta T_s + \Delta T_c + \Delta T_k \quad (2.16)$$

where T_0 is a convenient reference temperature, ΔT_s, ΔT_c, and ΔT_k are the undercoolings resulting from solute, curvature, and kinetics, respectively. The kinetic undercooling ΔT_k arises because of non-equilibrium effects; this has been discussed in the previous section. The curvature undercooling ΔT_c represents the depression in freezing temperature due to the Gibbs-Thompson curvature effect. For surfaces whose growth orientation does not coincide with a cusp in the Wulff plot [21],

$$\Delta T_c = \frac{\gamma}{\Delta S}\left[\left(\frac{1}{r_1} + \frac{1}{r_2}\right) + \frac{\partial^2 \gamma}{\partial n_x^2}\frac{1}{r_1} + \frac{\partial^2 \gamma}{\partial n_y^2}\frac{1}{r_2}\right] \quad (2.17)$$

while for faceted or singular orientations (*i.e.*, at a cusp in the Wulff plot), the curvature undercooling term becomes [22]

$$\Delta T_c = \frac{1}{\Delta S\, A}\int (\csc\theta - \cot\theta)\, dL \quad (2.18)$$

The terms in this equation are defined in Fig. 2.12 and the integration is performed over the perimeter of the facet. For isotropic materials, both equations reduce to the form used by Mullins and Sekerka [23]

$$\Delta T_c = A\left(\frac{1}{r_1} + \frac{1}{r_2}\right) \quad (2.19)$$

where A is the Gibbs-Thompson coefficient, r_1 and r_2 are the principal radii of curvature.

If T_0 is the melting temperature for an alloy of composition C_0, the solute undercooling ΔT_s is given by

$$\Delta T_s = m (C_0 - C_{LI}) \qquad (2.20)$$

where m is the liquidus slope. The local liquid composition C_{LI} departs from the bulk composition because of solute redistribution at the interface,

$$C_{SI} = k\, C_{LI} \qquad (2.21)$$

where k is the equilibrium solute distribution coefficient, C_{SI} is the solid solute concentration at the interface. During fast growth, such as in the laser surface remelting of alloys [24], the usual assumption of local equilibrium at the solid/liquid interface is no longer valid, and k deviates from the equilibrium value, i.e., k = f(R), leading to solute trapping at high growth rates [25-28].

The solute and heat flow at the interface are described [29] by:

$$R\,(C_{SI} - C_{LI}) = D_L \frac{\partial C}{\partial n} - D_S \frac{\partial C}{\partial n} \qquad (2.22)$$

$$R\,L = K_S \frac{\partial T}{\partial n} - K_L \frac{\partial T}{\partial n} \qquad (2.23)$$

where R is the growth rate, n is the normal to the interface, D_L and D_S are the liquid and solid diffusion coefficients, L is the latent heat per unit volume, K_L and K_S are the liquid and solid heat conductivities.

If k < 1, solute is rejected at the interface and thus enriched in the liquid ahead of the interface. If k > 1, on the other hand, solute is depleted in the liquid ahead of the interface.

2.2.2 Heat and solute transport

Neglecting fluid motion, heat and solute flow occurs by conduction and diffusion [30]. The equations:

$$D_{iT} \nabla^2 T = \frac{\partial T}{\partial t} \qquad (2.24)$$

$$D_{iC} \nabla^2 C = \frac{\partial C}{\partial t} \qquad (2.25)$$

must be solved in the solid and liquid phases (t is time, D_{iT} and D_{iC} are the thermal and solute diffusivities in the relevant phases).

For steady state growth, these equations may be transformed to coordinates moving with the interface, and become [31]

$$D_{iT} \nabla^2 T + R \frac{\partial T}{\partial x} = 0 \qquad (2.26)$$

$$D_{iC} \nabla^2 C + R \frac{\partial C}{\partial x} = 0 \qquad (2.27)$$

where x is the steady state growth direction.

In fact, fluid flow occurs in most solidification processes, and the effect of convection should be included in the transport equations [29,32-34].

2.2.3 Planar front stability

2.2.3.1 Constitutional undercooling

The instability of a planar interface, *i.e.* the breakdown of a planar interface into cells, was first explained by Tiller *et al.* [31] in terms of constitutional undercooling. It was predicted that cells should form when

$$G_L < m\, G_C \qquad (2.28)$$

where G_L is the temperature gradient and G_C is the concentration gradient in the liquid.

2.2.3.2 Mullins-Sekerka theory

Mullins and Sekerka [23] put the stability criterion on a firmer theoretical basis by considering the condition when the amplitude of an infinitesimal sinusoidal perturbation in the originally planar interface shape just began to grow. The inequality at low growth rates then becomes

$$G' < m\, G_C \qquad (2.29)$$

with

$$G' = \frac{K_S G_S + K_L G_L}{K_S + K_L} \qquad (2.30)$$

where G_S is the temperature gradient in the solid. When $G_S = G_L$ and $K_S = K_L$ is assumed, this equation becomes identical with the constitutional undercooling criterion. This treatment however has provided new insights into the stability problem.

The development of shallow cells from a planar interface has recently been modelled numerically [35-39].

2.2.4 The growth condition

2.2.4.1 Isolated dendritic growth

When a solid is nucleated in the centre of an undercooled melt, a negative temperature gradient exists in front of the interface because of the release of latent heat, and dendrites grow from the nucleus into an infinite bath at an approximately constant growth rate determined by the bath undercooling.

2.2.4.2 Array growth

When alloys are grown directionally, for example by pulling a specimen out of a furnace at a constant rate, a positive temperature gradient exists in front of the interface. After the breakdown of the planar interface, arrays of cells or dendrites are formed. The spacing of cells/dendrites formed depends on the growth rate, temperature gradient and alloy composition.

2.2.5 Steady state non-faceted cellular array growth

Work in the past has been concentrated on non-faceted growth, which is mostly found in metals. Because of the difficulty in carrying out time-dependent analysis, theoretical treatments normally assume steady state growth. A number of approximate steady state models have been proposed for array cells. Burden and Hunt [40-42] assumed that the cell tip was part of a sphere, and attempted to allow for the interaction of the diffusion fields between neighbouring cells using an approach suggested by Bower *et al.* [43], and Laxmanan [44] has extended this approach. Kurz and Fisher [45] assumed that the cell tip was elliptical and approximated the diffusion solution to that for an isolated dendrite. Trivedi [46] modified the isolated dendrite model to include a positive temperature gradient. The latter Refs. [45,46] however do not make an allowance for the overlapping of the diffusion fields from neighbouring cells.

All of these models may be criticized because they arbitrarily assume a cell tip shape. A numerical method has recently been used by Hunt and McCartney [47,48] to calculate truly self-consistent three dimensional cell shapes. Only one cell shape was found for a fixed spacing, velocity, temperature gradient and composition; however solutions were found to be present over a wide range of spacings for a fixed growth condition, as was found by approximate analytical models. However, practically only a narrow range of spacings are found for any given growth condition. Clearly some other condition must be specified to define cellular growth. The same situation is also true of eutectic growth and dendritic growth.

2.2.6 Stability treatment

In the past the objective of the stability treatment has been to predict a unique cell spacing under a given growth condition.

2.2.6.1 Extremum growth hypothesis

It has been suggested that [40-42,49-51] cells or dendrites grew at the minimum undercooling for a given growth velocity, which is equivalent to the maximum growth rate at a given temperature; *i.e.*, the structure approaches optimum growth. The assumption has been shown to hold reasonably well with eutectic growth [20,52,53]. However careful comparison with experimental work has shown that it does not correctly describe cellular growth [48]. In fact no physical justification has ever been cited for this *ad hoc* assumption.

2.2.6.2 Marginal stability hypothesis

Burden [40] first applied the Mullins-Sekerka type technique to a cell tip in isolation to calculate the largest tip radius that was just stable to a tip splitting perturbation. A similar approach was later adopted independently by Langer *et al.* [54-56] for isolated dendritic growth; and the stability has been termed 'marginal stability'. Other workers [45,46] have also made identical prediction to that of Burden; *i.e.*, a cell tip is marginally stable when the tip radius is less than

$$r = \sqrt{\frac{A\eta}{m\,G_C - G}} \qquad (2.31)$$

where $\eta = 4\pi^2$.

Careful experimental work [57] has shown that the tip radius of an isolated dendrite is correctly predicted by the marginal stability condition. Comparison with experiments for array growth has indicated [48] that the marginal stability condition does not describe array cellular growth correctly. As has been shown by the numerical work [47,48], the proximity of neighbouring cells affects the stability quite considerably; this however has been neglected in the marginal stability treatment.

2.2.7 Experimental work

Experimental work on steady state non-faceted cellular array growth includes the measurement of cell spacings, λ, and cell tip temperatures and compositions. These include measurements in metals [59-61,64-67] as well as in non-faceting transparent organic compounds [62,63,68-74]. The results however appear to show contradictory behaviour. Venugopalan and Kirkaldy [63] reported that no steady state cellular spacings were observed near V_c, the critical velocity for the planar interface stability, in the succinonitrile-salol system. Sharp and Hellawell [64] studied the variation in cell spacing with velocity in an Al-Cu system by changing the velocity in steps. They found that the cell spacing did not change appreciably as the velocity was increased near the critical velocity. By using the decantation technique, Rutter and Chalmers [65] and Tiller and Rutter [66] found that cell spacings decreased with an increase in velocity. A similar result was qualitatively reported by Jin and Purdy [67] in Fe-Ni alloys. Cheveigne *et al.* [68] and Venugopalan and Kirkaldy [63] studied the variation in cellular spacing with velocity in transparent organic systems and they found a similar behaviour at velocities larger than V_c. Trivedi *et al.* [69-72] and Esaka and Kurz [73] found cell spacings to increase with velocity near the cell-dendrite transition velocity in the succinonitrile-acetone system. A similar variation in cell spacing was also observed by Bechhofer and Libchaber [74] in an impure pivalic acid system.

It should be pointed out that, experiments with transparent organic compounds in the form of thin films are not comparable with three dimensional cellular growth found in castings and treated in typical theoretical models (see Chapters 4 & 5). Vinals et al. [75] investigated the effect of the finite sample thickness on the stability of the solidifying front, and concluded that the onset of instability of a planar interface depends on the thickness of the sample.

In the more recent work of McCartney and Hunt [59,60] with aluminum alloys, efforts were made to reduce the effect of convection. Their results gave

$$\lambda \propto G^{-0.5} V^{-0.25} C_0^{0.3} \tag{2.32}$$

2.2.8 Dendritic growth

Dendrites form further away from the planar front stability condition. Isolated dendritic growth is very different from array growth. The characteristics of isolated dendrites are the bath undercooling and the tip radius [76-81]. Arrayed dendrites have been approximated by a smooth cell-like shape [40-48]. However the real dendritic shape is not smooth, and the cell spacing and tip radius may vary separately. The side-branching behaviour of dendrites are in fact significantly time-dependent [82].

2.2.9 Faceted cellular array growth

Faceted cellular growth occurs when a planar front of a faceting crystal breaks down. Bardsley et al. [83-86] and Cockayne [87] studied the development of faceted cells in germanium alloys and other semiconductors and melt-grown oxides. They have shown that germanium single crystals break down to form cells as predicted by the constitutional undercooling criterion, i.e., at a very small amount of constitutional undercooling. They have also found that deep cell boundary grooves form if sufficient solute is present. Tarshis and Tiller [88], O'Hara et al. [21], and Chernov [89] studied the effect of kinetics on the stability of facets, and concluded that the kinetics of faceted growth had stabilizing effects.

More recent work on thin film silicon single crystals (Silicon-on-Insulator, or SOI) produced by zone melting recrystallization has provided ample evidence of faceted cellular growth in silicon. These thin film silicon single crystals are produced for solar cells and integrated circuits. An important feature of the work in this area has been the presence of cell boundary networks [28,90] (Fig. 2.13), regardless of the heating methods used. These defects can reduce the performance of the semiconductor device greatly.

Although there have not been sufficient consistent quantitative results available, experimental work has shown some general features: (1) The relationship between the cell spacing and the growth velocity is more complex than non-faceted cellular growth. For example, Geis et al. [91] reported that the cell spacing increases with growth velocity, approximately as the square root, and Im et al. [92] found that the cell spacing increases with

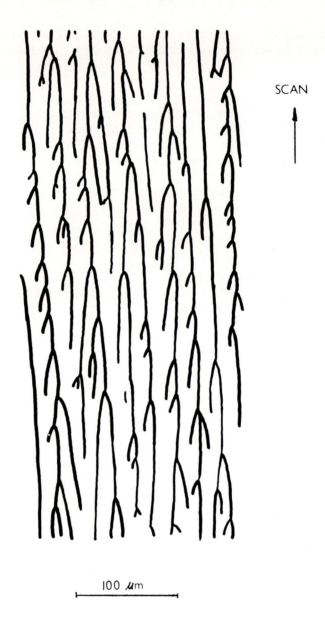

Fig. 2.13 Nomarski micrographs of typical subboundary networks in recrystallized thin film Si (from Ref. 90).

velocity over the low velocity regime and decreases with velocity over the high velocity regime. (2) The cell spacing decreases with temperature gradient [93-96]; (3) More irregular cell structures are associated with smaller cell spacings [93-96]; and (4) there are impurities present, and they are enriched along cell boundaries [97-98]. These impurities have been identified to be oxygen, carbon and nitrogen; however their thermodynamical behaviour in silicon is not clear, for example, the distribution coefficient of oxygen in silicon has been reported [99] to vary between 0.2 - 13. Other factors, *e.g.*, the reflectivity change with temperature, film thickness fluctuations, and thermal stresses, should also be considered [95,100].

Clearly faceted cells behave very much differently from non-faceted cells. Attempts have been made to explain the pattern formation in silicon, among which is the approach proposed by Pfeiffer *et al.* [90], which explains the pattern formation in silicon in terms of facets interactions. The details of this model will be reviewed later (see Chapters 8 & 9). Although significant criticisms can be made on this model, it does appear to be an approach in the right direction.

The lack of understanding in faceted cellular growth has been the result of this area being ignored in the past. However, faceted cellular growth plays a vital role in the crystal growth of a number of important semiconductor materials, including Si, Ge, *etc.*

2.2.10 Pattern formation

Cellular growth, faceted and non-faceted, dendritic growth and eutectic growth all offer excellent examples of pattern formation in nature. The question to be answered is: how nature selects the pattern?

Experimental evidence has suggested that these pattern formation processes are intrinsically time-dependent. The pattern formation process in faceted cellular growth of silicon in particular has suggested strong cellular interactions in the array; the dynamical feature of the system should play a vital role in pattern formation. Recent theoretical work by Datye *et al.* [101] treated the time-dependent eutectic growth in a slightly noisy environment. Karma [102] constructed a random-walk model to simulate the time-dependent eutectic solidification. These models demonstrated the complex dynamical behaviours of eutectic growth.

It can also be anticipated that such factors as convection and crystalline anisotropy may play important roles in pattern formation.

2.3 Objective of the Present Work

The main objective of this work has been to observe the faceted cellular growth process *in situ*, and to model the time evolution of the faceted cellular interface, in an attempt to address the pattern formation problem. Work has also been carried out to study steady state non-faceted cellular array growth.

Chapter III

INTRODUCTION TO NUMERICAL MODELLING

3.1 Numerical Solution of Differential Equations

The technique of computer modelling has found a wide variety of applications in the study of solidification, ranging from microscopic to macroscopic phenomena. As has been mentioned in the previous chapter, during the solidification process, the growth of the interface is governed by heat and solute transport, and related thermodynamical and kinetic laws. Computer simulation of the growth process means solving numerically the differential equations for heat and solute flow subject to appropriate boundary conditions, coupled with growth kinetics at the interface.

In order to solve a differential equation numerically for a dependent variable ϕ (temperature, composition, *etc.*), the continuous distribution of ϕ described by the equation is replaced by a set of discrete values associated with a finite number of locations, called grid-points; *i.e.*, the distribution of ϕ is discretised. By assuming some approximate piecewise algebraic profile for ϕ in its vicinity, the governing differential equation is then replaced by a set of simple algebraic equations expressing the variation of ϕ in terms of the grid-point values; these are the discretisation equations. The discretisation equations, if set up properly, should describe the same physical information as the differential equation.

3.2 Finite Difference and Finite Element Methods

With the finite difference method, the discretisation equations are obtained by considering the differential equation only at the grid-points, and the solution consists solely of a set of values for ϕ at these specific locations. With the finite element method, on the other hand, the discretisation equations are derived by considering the variation of ϕ described by the differential equation, throughout the whole region of interest and not only at a finite number of points, and the solution comprises a set of grid-point values and a set of interpolating functions.

3.3 The Formulation of Discretisation Equations

There are a number of different ways of deriving the discretisation equations, including the Taylor series formulation, the variational formulation, the weighted residual formulation, and the control volume formulation. The method used in this work is the control volume finite difference method. With this method, the solution domain is divided into a set of non-overlapping sub-domains, called control volumes, and the differential equation is integrated over each of these separately to yield the discretisation equations; in this integration, ϕ is approximated by piecewise profiles. Each discretisation contains the value of ϕ for a group of grid-points, and is an expression of the conservation principle for the corresponding control volume, just as the differential equation is an expression of the conservation principle for an infinitesimal volume. Thus each discretisation equation can be obtained by applying directly the conservation principle to a control volume after having specified fluxes at its faces. Since the bulk property corresponding to ϕ is conserved in each control volume individually, it is also conserved throughout the entire solution domain.

3.4 Grids

As shown in Fig. 3.1, the solution domain $0 \leq x \leq L$ is divided into an array (uniform or non-uniform) of control volumes. There are two conventions for positioning grid-points relative to the faces of the control volumes: (1) Place the faces of the control volumes midway between the grid-points (Fig. 3.2(1)); or (2) place the grid-points midway between the faces (Fig. 3.2(2)). The second scheme is normally favoured, because the grid-point value can be considered to be a reasonable average for the control volume, and the flux calculated at any face is representative of the entire face.

3.5 Interpolation (Extrapolation)

Interpolation is used to calculate the value of ϕ at the faces of the control volume and the gradient across the faces needed for the calculation of flux terms across the faces, from the values at neighbouring grid-points. To do this, a piecewise profile needs to be assumed for ϕ. The most convenient profile is a linear one, as shown in Fig. 3.3; other profiles may be preferred if they can give a better accuracy.

3.6 The Steady State Problem and Time-Dependent Problem

For a steady state problem, the variable ϕ does not vary with time, *i.e.*, $\frac{\partial \phi}{\partial t} = 0$. Compared with the steady state problem, the inclusion of a time dependency for a time-dependent problem is not difficult to accommodate conceptually; time just represents another dimension to the solution domain and therefore a further dimension in the array of grid-points.

There are three different schemes (Fig. 3.4) for the piecewise profile of ϕ with time t during the time step δt:

1. The explicit scheme, which assumes that the value of ϕ at the current time, *i.e.*, the old value $\phi(t)$, remains right up until the next time $(t + \delta t)$, when it changes abruptly to its new

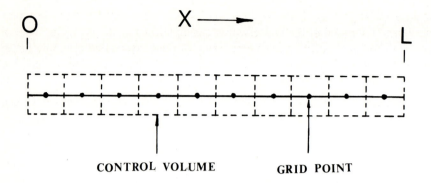

Fig. 3.1 The array of control volumes.

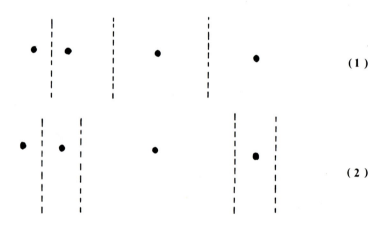

Fig. 3.2 The two conventions for positioning grid-points.

(1) Control volume faces midway between points; and

(2) points midway between faces.

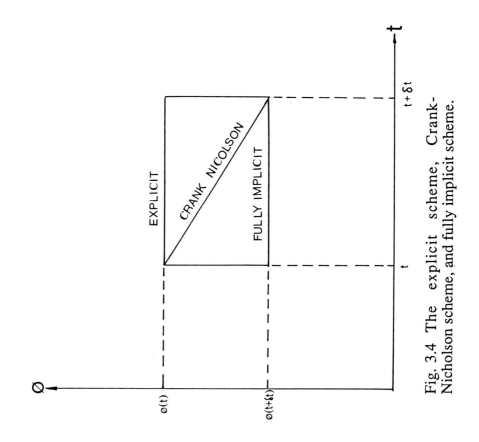

Fig. 3.4 The explicit scheme, Crank-Nicholson scheme, and fully implicit scheme.

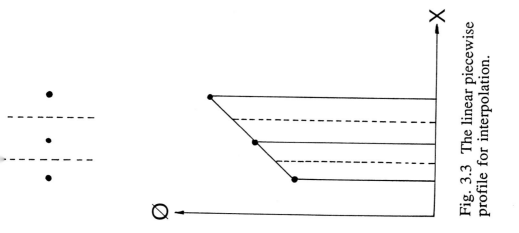

Fig. 3.3 The linear piecewise profile for interpolation.

value, $\phi(t + \delta t)$. With this scheme the solution can be obtained by marching in time from a given initial value of ϕ.

2. Crank-Nicholson scheme, which assumes that ϕ varies linearly with t from $\phi(t)$ at t, to $\phi(t + \delta t)$ at $(t + \delta t)$. This scheme results in the discretisation equations consisting of known, old values $\phi(t)$ and unknown values $\phi(t + \delta t)$.

3. Fully implicit scheme, which assumes that ϕ attains its new value $\phi(t + \delta t)$ at the very start of the time step. This scheme results in the discretisation equations consisting of unknown values $\phi(t + \delta t)$.

Unlike the first scheme, for the last two schemes, the set of discretisation equations obtained need to be solved in order to obtain the new values $\phi(t + \delta t)$.

3.7 Convergency, Consistency and Stability

3.7.1 Convergency

If U is the exact solution of a partial differential equation and u is the exact solution of the corresponding discretisation equations, then the discretisation equations are said to be convergent when u tends to U as the grid is refined. The quantity (U - u) is called the discretisation error, and it depends on the discretisation intervals and the order and nature of the piecewise profiles assumed to describe the dependent variable ϕ.

3.7.2 Consistency

If the local truncation error at the grid-point, which is brought in on deriving the difference equations, tends to zero as the discretisation intervals tend to zero, then the discretisation equations are said to be consistent with the partial differential equation. It is possible for a set of discretisation equations to be consistent with two partial differential equations.

3.7.3 Stability

Stability is only relevant when the problem is time-dependent. The total error in the numerical solution consists of the discretisation error, as described above, and the rounding error in the computer, although in modern computers, the rounding error is invariably smaller than the discretisation error. The numerical solution is said to be stable if it is bounded; *i.e.*, there is an upper limit, as $\delta t \to 0$, to the extent to which any piece of information can be amplified in the computations. This requires both the discretisation error and the rounding error to be bounded. For linear differential equations, consistency and stability guarantee convergency.

The Crank-Nicholson and fully implicit schemes are unconditionally stable over time. Oscillations can set in with the Crank-Nicholson scheme, however, and physically unrealistic solutions can be obtained unless very small time steps are used. The fully implicit scheme however can always ensure physically realistic solutions. The explicit scheme is stable when:

$$\delta t \leq \frac{\delta x^2}{2D}$$

where δx is the size of the control volume (for one dimensional domain), and D is the relevant diffusivity.

Practically, during the calculation the numerical solution can be monitored, and if the solution is evolving in an apparently controlled and physically realistic manner, it is assumed to be stable. If reducing the grid sizes has little effect on the solution, it is assumed that it has converged sufficiently close to the exact solution.

3.8 Other Numerical Methods

Other numerical methods used in this work include numerical integration, the Newton-Raphson method for solving transcendental equations, and the Gaussian elimination method or Gauss-Seidel iterative method for solving a set of linear equations. Descriptions of these methods can be found in standard textbooks of applied mathematics.

Chapter IV

EXPERIMENTAL TECHNIQUES AND APPARATUS

4.1 Introduction

The purpose of the experimental work was to observe faceted cellular growth *in situ* in order to study the physical process, and to study the interface shape for steady state non-faceted cellular array growth. Jackson and Hunt [4] first realized that some aspects of the solidification of metals can be studied using some transparent organic compounds as analogues, as they exhibit similar solidification behaviours. This has indeed in the past helped greatly to provide a deep insight into solidification processes. This technique is used in the present experimental work. The experimental efforts include: (1) Construction of a temperature gradient stage (TGS) to provide the growth condition; (2) Selection and purification of materials; (3) Preparation of the specimen; and (4) Crystal growth and observation.

4.2 The Temperature Gradient Stage

4.2.1 General considerations

The TGS was required to provide highly stable and accurately adjustable growth conditions, *i.e.* growth velocities and temperature gradients, for directional solidification. The growth of the interface was to be observed directly under the optical microscope.

4.2.2 The design and construction

4.2.2.1 The main body

The main body of the TGS consisted of a water-cooled cold-plate and a hot-plate which was kept at a constant temperature. Copper was chosen as the material for the plates because of its good thermal conductivity, so that a uniform temperature can be readily achieved. Figs. 4.1 & 4.5 show a photograph and a schematic diagram of the TGS built. The hot-plate and the cold-plate are supported by the Teflon supporter. Above the plates is a trolley made of Teflon, which can travel along the Teflon track groove very smoothly. Teflon was chosen as the material for the track and the trolley because it has very little friction. The Teflon supporter was then mounted onto a water cooled brass base-plate.

4.2.2.2 Motion of the specimen

The specimen placed on the plates was pulled by the trolley which was driven at a set speed by a variable-speed DC motor through a pair of 1:1 gears and a single thread drive rod. The D.C motor was powered by a Variac. Motors available included an MD83 (General Time) motor with a nominal rotational rate of 1 RPM at 12 volts, and a DME 32S motor with a nominal rotation rate of 5500 RPM plus a reduction gearbox giving a rotation rate of about 10

Fig. 4.1 The temperature gradient stage.

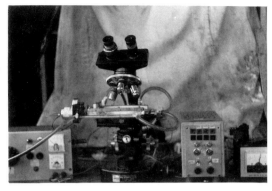

Fig. 4.2 The distiller.

Fig. 4.3 The zone-refiner (horizontal position).

Fig. 4.4 Zone-refining in operation.

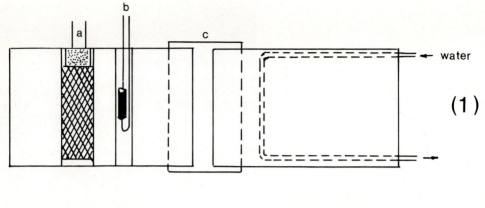
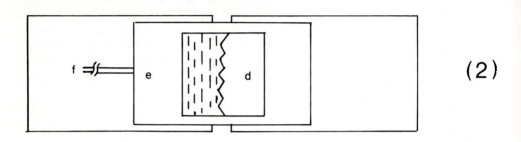

Fig. 4.5 Schematic diagram of the temperature gradient stage.

(1) The bottom plate, and

(2) the top plate.

a: Solon heater; b: Thermistor; c: Cover glass;

d: Glass cell; e: Teflon trolley; f: Drive rod.

RPM at 12 volts. The surface of the plates were made very smooth so that the specimen can move on the plates smoothly and in constantly good contact with the plates.

4.2.2.3 Heating

Heating of the hot-plate was provided by a 25W Solon soldering iron heating element, the output of which was controlled by an electronic controller. The heating element was securely sandwiched between the upper and lower heating plates to ensure maximum heat transfer. The input to the temperature controller was provided by a 10K resistance-temperature curve matched thermistor with a 2.4 mm bead, which was sheathed in plastics to ensure electrical insulation from the hot-plate, and accommodated by a shallow recess milled in the plate. The input to the heater was simultaneously read with a digital multimeter to provide information about the heating process.

4.2.2.4 The stage

The TGS was mounted onto the stage of an transmission optical microscope, and it can be translated on the microscope stage in the plane normal to the axis of illumination which passes through the specimen above the gap between the hot-plate and the cold-plate. The gap was sealed by a glass plate underneath the specimen to achieve double-glazing effect. The specimen was then capped with a glass cap to eliminate thermal fluctuations. The cap also serves to press the specimen onto the plates.

4.2.3 Calibrations

The main growth parameters, *i.e.* the moving rate of the specimen and the temperature gradient, need to be calibrated.

4.2.3.1 The moving rate of the specimen

The moving velocity of the specimen was calibrated by dividing the distance a spot on the specimen travelled, read off from the graticule scale of a 10x eyepiece and calibrated with a micrometer, by the time taken for the movement read with a Trackmaster stopwatch. For each calibration this was repeated 4 times and the average was taken. For each motor this was done for a number of discrete input voltages applied. The results were summarized in Table 4.1.

4.2.3.2 The temperature gradient

The temperature gradient was produced by the temperature difference between the water-cooled cold-plate and the hot-plate heated to a constant temperature. The temperature gradient can be adjusted by changing the width of the gap between the two plates. The calibration was carried out at the beginning of each set of experiments, during which both the plate temperature and the gap width was kept constant giving a constant temperature gradient. The temperature gradient was measured by measuring the temperatures at two points in the melt in the specimen (for specimen preparation see 4.4.2) and dividing their difference by their distance along the temperature gradient. The temperature was measured by 100μm Chromel-Alumel thermocouples soldered with silver solder flux over a flame producing a joint of 200μm in size.

Table 4.1 Calibration of the Moving Velocity of the Specimen

Motor	Applied Voltage (V)	Velocity (μm/sec)
General Times (1 RPM)	2.2	1.1
	4.2	2.5
	5.0	3.3
	6.5	3.9
	8.5	5.7
	9.0	6.1
	9.5	6.4
	10.5	6.8
DME (∼10 RPM)	1.3	8.6
	2.3	15.1
	4.2	28.3
	5.3	35.5
	6.5	43.4
	9.0	60.2

The thermocouples were introduced into the melt parallel to the isotherm (*i.e.* perpendicular to the temperature gradient) thus minimizing their disturbance to the thermal field (Fig. 4.6). The two thermocouple joints were placed in the same line along the temperature gradient, and their distance was measured on the graticule scale calibrated with a micrometer. The thermocouple leads went to an ice-water cold junction where they were twisted to screened copper cables which were taken to a Kipp and Zonen BD101 6-channel chart recorder. The cold junction was schematically shown in Fig. 4.7. The twisted junction was held at the bottom of a glass tube filled with ethanol. The voltage from the thermocouple was then converted into temperature. The measurement device (Fig. 4.8) was left on overnight and values were taken when the thermal field had become stabilized.

4.2.4 Operation and performance

Results from the temperature gradient measurements suggested that the temperature field obtained in the specimen was reasonably stable (This was also evidenced by the observation that the planar solid/liquid interface at equilibrium remained stationary over the period of several hours). Results of repeated measurements of the specimen moving rate indicated that the moving rate was constant over the whole distance of the specimen movement. In an experimental run, about 5 hours was spent for the thermal field to stabilize. This was indicated by the constant power input into the heating element, and confirmed by the stable, planar interface in the specimen. Water cooling was supplied from the beginning. Brightness of the lighting source of the microscope was kept constant during a run.

4.3 Material Selection and Purification

4.3.1 Material selection

The materials to be used are organic compounds which are either faceting or non-faceting depending on the purpose of the experiment. The other criteria are: (1) optically transparent; (2) melting points between room temperature and 100^0C so that it can be easily handled. Some of these commonly used materials and their properties are listed in Table 4.2. Materials used in this work included salol (phenyl salicylate), thymol, and *o*-terphenyl for faceted growth, and succinonitrile for non-faceted growth. Stock materials available included 99% salol, 98% thymol, 99% *o*-terphenyl, 97% succinonitrile, supplied by Aldrich Chemical Co. Ltd, and 99.5% Acetone supplied by BDH Ltd.

4.3.2 Material purification

Material purification is necessary because impurities can have very significant effects on solidification. Materials used in this work were purified first by distillation and then by zone refining.

4.3.2.1 Vacuum distillation

It was first attempted to purify salol by distillation. The apparatus constructed were shown in Figs. 4.2 & 4.9. All the test tubes and extension tubes were made of Pyrex glass. The

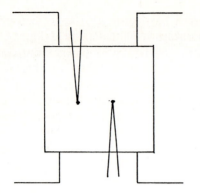

Fig. 4.6.

Schematic diagram showing the positions of the thermocouples.

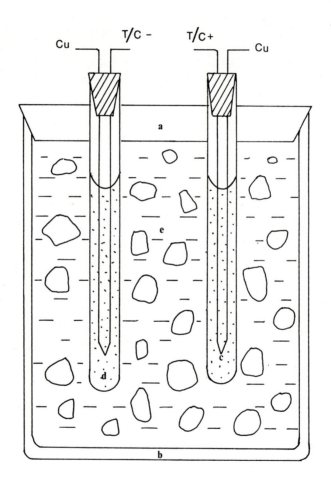

Fig. 4.7 The cold junction.

a: Cork; b: Dewar; c: Junction;
d: Ethanol; e: Ice water mixture.

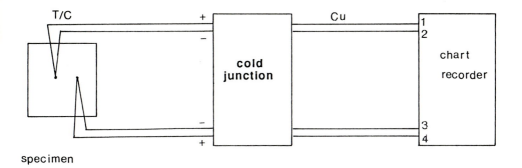

Fig. 4.8 The measurement circuit.

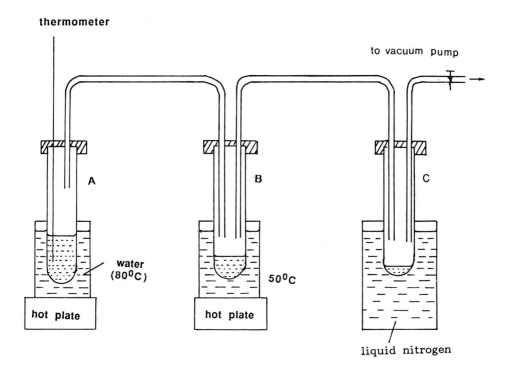

Fig. 4.9 Schematic diagram of the distiller.

Table 4.2 Properties of Organic Compounds Used

M.P: Melting point (°C);
B.P: Boiling point (°C) at 760 mm Hg pressure, unless otherwise specified;
α : Jackson's α factor;
[ ]: References.

Substance	Structural Formula & Molecular Form	Crystalline Form	M.P	B.P	α	Toxicity
Phenyl Salicylate (Salol)	2-(HO)$C_6H_4CO_2C_6H_5$ [103] [103]		43 [106]	172-173 (12 mm) [103]	7 [3]	Toxic [103]
Thymol	2-[($CH_3)_2$CH]C_6H_3 5(CH_3)OH [103] [107]	Plates [107]	51.5 [111]	232 [103]	6.5 [104]	Irritant [103]
o-Terphenyl	$C_6H_5C_6H_4C_6H_5$ [103] [106]	Monoclinic prisms [106]	58-59 [103]	337 [103]	6.2 [105]	Irritant [103]
Succinonitrile	$NCCH_2CH_2CN$ [107]		58.08 [57]	265-267 [103]	1.4 [3]	Toxic Irritant [103]

junctions were sealed with vacuum grease. The glassware was cleaned with acetone and dried before use. Water was used as heating medium since its temperature cannot significantly exceed 100^0C, and so there should be no danger of overheating the chemicals. Heating was provided by hot plates. The whole device was clamped onto a metal stand and placed in a fume cupboard. Tube A contained stock materials, and the water bath temperature was controlled at 80^0C. Material species with boiling points below 80^0C (under vacuum) were evaporated and led into tube B, which was kept at 50^0C. Gaseous species continued to flow into tube C, leaving salol condensed into liquid drops and deposited in tube B. Tube C was cooled with liquid nitrogen, and gaseous species were trapped there to prevent them from contaminating the vacuum pump. However, this simple distillation technique was later found not to be effective enough, as the distilled material still contained a lot of impurities which formed gas bubbles in the specimen. Some more powerful purification technique needed to be employed.

4.3.2.2 Zone refining

(1) Introduction

Due to the different solubilities of a solute species in the solid and the liquid phase of its solvent, the solute content is either enriched (if k < 1) or depleted (if k > 1) in the liquid ahead of the interface during solidification. Thus each time a molten zone travels through a portion of solute containing solid, the solute will be redistributed. This principle led to the discovery of zone melting by Pfann in 1952 [109]. Zone refining is by far one of the most powerful techniques for material purification.

In this work, following the discovery of the incapability of the distillation technique for material purification, the zone refining technique was resorted. A zone refiner was constructed with which all the materials used were purified.

(2) Principles of zone refining [110]

(2-i) Single pass distribution

As shown in Fig. 4.10(1), after a molten zone of length d passes through an ingot of length ℓ with an originally uniform composition C_0, the solute distribution (assuming k < 1) is approximately as shown in Fig. 4.10(2). The curve has three distinct regions: an initial region, a level region, and a final region. For the first two regions, the solute distribution can be described by the equation:

$$\frac{C}{C_0} = 1 - (1 - k) e^{-kx/d} \qquad (4.1)$$

where C is the solute concentration in the solid, and k is the distribution coefficient. Fig. 4.10(3) shows the solute distribution in the solid for various values of k. As can be seen, the concentration in the initial region is less the C_0. So this is a region of purification.

(2-ii) Multipass distribution

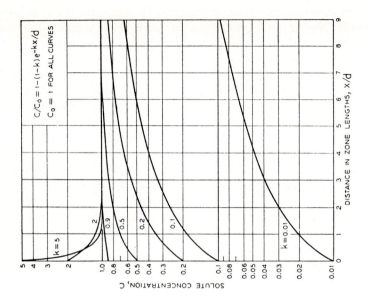

Fig. 4.10(3) Curves for single-pass zone melting showing solute concentration in the solid versus distance in zone lengths from beginning of charge, for various values of the distribution coefficient k (after Ref. 110).

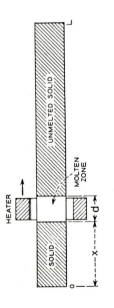

Fig. 4.10(1) Molten zone of length d traversing a cylindrical ingot of length L (after Ref. 110).

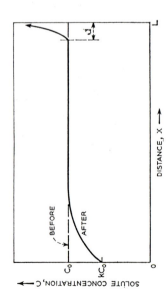

Fig. 4.10(2) Approximate solute concentration after passage of one molten zone through a charge of uniform mean concentration C_0 (after Ref. 110).

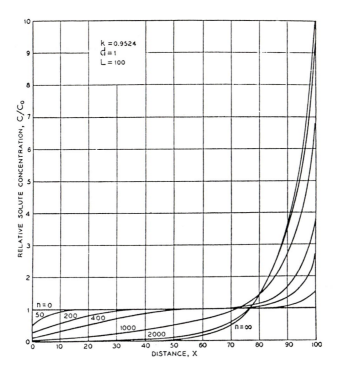

Fig. 4.10 (4)

Relative solute concentration C/C_0 versus distance x with number of passes n as a parameter for k = 0.9524, d = 1, L = 100 (after Ref. 110).

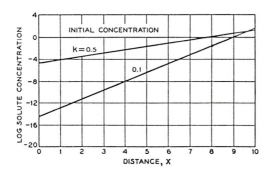

Fig. 4.10 (5) Ultimate distribution attainable by zone refining, for k = 0.5 and k = 0.1, for an ingot 10 zone lengths long (after Ref. 110).

As the process described above is repeated, the length of the purification region increases. Fig. 4.10(4) shows the relative solute concentration with number of passes as a parameter.

(2-iii) The ultimate distribution

After many passes, the solute distribution approaches a steady-state, or ultimate, distribution that represents the maximum attainable purification, as shown in Fig. 4.10(5).

(2-iv) Efficiency of zone refining purification

Reducing the length of the molten zone reduces the length of the final region thus increasing the efficiency of purification. Increasing the diffusion of solute will increase the effective distribution coefficient and therefore increase the efficiency. A non-planar solid/liquid interface will give rise to inhomogeneity across the cross section. So a planar interface is more favourable for achieving uniform purification.

(3) Construction of the zone-refiner

(3-i) The requirement

The requirement for the zone-refiner was to produce a molten zone in the material which can travel unattended at a set speed repeatedly over a certain distance. This would in turn require a heater which can heat the material to be zone-refined above its melting temperature; a trolley to carry the heater to move smoothly; a motor electronically controlled to facilitate the designed motion of the trolley, which is to move forward slowly over a distance, backward rapidly to the starting point, and start again.

(3-ii) The heater

The heater was a coil made of T1 Alloy resistance wire 0.35mm in diameter. The width and diameter of the coil should be small in order to form a narrow molten zone. The diameter of the coil is just large enough to allow the glass column to slip through smoothly. The heater was powered by a Wild microscope lighting power supply unit (MTR 12). The power to the heater was just great enough to produce a narrow completely molten zone. Caution must be taken against the short-circuiting of the coil. The heater was then fixed firmly on the trolley.

(3-iii) Motion of the trolley

The trolley was driven by an EC Motomatic Motor Generator through an adaptor, a gear box, and a single thread drive rod (Fig. 4.11). The whole system was carefully aligned to ensure smooth drive. The revolution of the motor was controlled by an E550 Motomatic speed control unit, so that the motor can provide rotations in two directions at different speeds. The changeover of the rotation direction was triggered by switches A and B through the operation of an electronic circuit (Fig. 4.12). The following movement can be achieved by the trolley: a slow movement towards switch B until it reaches B, when it moves back rapidly until it reaches switch A, where the cycle of movement starts again. The circuit consists of the control unit,

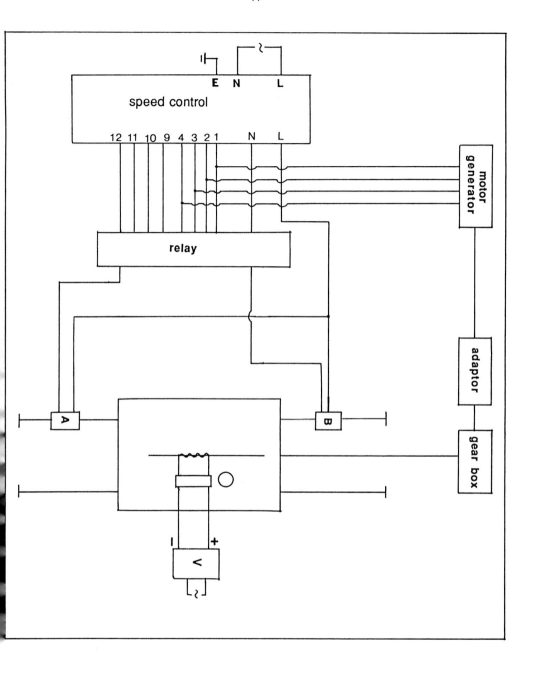

Fig. 4.11 Panel diagram of the zone-refiner.

A, B: Microswitches.

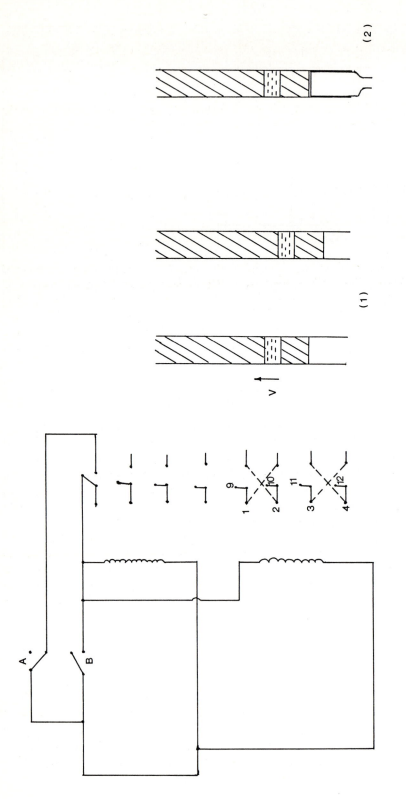

Fig. 4.12 The circuit which controls the movement of the trolley. A, B: Microswitches.

Fig. 4.13 (1) Showing the immigration of the sample along the tube. (2) An improved design.

two 115 V AC relays with 4-pole changeover, and two SPDT momentary action microswitches. The normal state of the switches is: A (on); B (off), which gives the connection as:

State 1: 1 / 9 ; 2 / 10 ; 3 / 11; 4 / 12

leading to the clockwise rotation of the motor. Once switch B is depressed, operation of the relays changes the connection into:

State 2: 1 / 10 ; 2 / 9 ; 3 / 12 ; 4 / 11

leading to the anticlockwise rotation of the motor, until switch A is depressed when the connection changes back to State 1.

(3-iv) The assembly

The trolley, motor, and controller were all assembled (Fig. 4.3) onto a wooden board supported by a number of rubber corks. The sample was clasped onto steel stands fixed on the board. The sample must be carefully aligned to be parallel to the axis of the drive rod. The whole assembly can either sit on a table or be hung along the wall. In the experiment it was hung along the wall so that the molten zone travelled upwards; the gravitational effect would help the lighter impurities, mostly gases, to float upwards. This will help increase the efficiency of zone refining.

(4) Preparation of the sample

A Pyrex glass tube was used as the container of the material for zone refining. The advantages of glass for this purpose are thought to be: (1) It is easy to be made into desired shape and size; (2) It is easy to clean, and has low reactivity and contaminativity; (3) It is transparent so that the crystal growth process can be observed through an optical microscope travelling with the trolley; and (4) the product can be taken out easily. The glass tube used was about 300 mm long and 4 mm in inner diameter. The tube need to be thin to provide good cooling condition, and to produce a planar interface in order to reduce inhomogeneity across the section. Since the length of the molten zone cannot be smaller than the tube inner diameter (as heat is penetrated about the same depth in all directions, see Chapter 6), a thin tube also favours the formation of a narrow molten zone. The glass tube was cleaned with acetone and dried. The material to be zone refined was first melted in a clean dish, and then filled into the glass tube slowly to allow air to get out. The tube was then cooled in the air before it was mounted onto the zone-refiner.

The original design was that the bottom of the tube was sealed and the tube was clamped onto steel stands. However the glass tube was always cracked after about 30 cycles. It was then realized that this resulted from the periodical stress produced by the expansion of the material on melting. Most faceting organic compounds have relatively large entropy of melting and expansion on melting. For example, the volume increase on melting for thymol is 7% of the volume of the solid at its melting point [111]. A modified design was to leave both ends of the tube open to release the stress. Another problem arose when it was noticed that the material down to the bottom of the bar immigrated at each cycle inside the tube against the moving direction of the molten zone (Fig. 4.13(1)), and consequently the last portion of material down to the bottom of the bar was zone refined for fewer cycles. An improved design was then made to overcome this problem. A portion of solid was always left at the bottom which was supported by a plastic bar (Fig. 4.13(2)). The plastic bar was just loosely fit with the tube, and was clasped onto the stands. The tube itself was supported by steel stands and no longer tightly fixed, so that it can move freely and little stress would build up. The expansion on melting was compensated by the movement of the tube in the opposite direction, so that the molten zone would always start from the same portion of material, in spite of the large expansion on melting. This design later proved to be very successful in overcoming the problems resulting from the large expansion on melting.

(5) Operation

To increase the efficiency of zone refining, it is desirable to have a narrow molten zone moving at a low velocity. The shape and size of the molten zone is determined by heat flow. A numerical model was developed to analyze the heat flow problem (see Chapter 6). During the zone refining process, the 1.7 ohm coil was powered at about 0.8 volts and moved at a velocity of 2.5 cm/hr, and a molten zone of 6 mm was formed. The zone refining was repeated for about 30-50 cycles. After zone refining the material was stored in a sealed dry test tube for later use.

The crystal growth process during zone melting was observed through an optical microscope. It was observed that during the process small gas bubbles were released from the freezing interface and flew upwards, and collected into bigger gas bubbles near the melting interface (Fig. 4.14(1)). The gas bubbles were finally transported to the top of the sample at the end of the cycle. As zone refining went on, less and less gas bubbles were observed in the molten zone. At the beginning of zone refining, a number of crystals were formed, and the solid looked obscure. As the material was purified, fewer crystals were formed at later runs, and the solid became clearer. Eventually a single crystal was produced (Fig. 4.14(2)) and the solid was clearly transparent (Fig. 4.4). This can be expected since impurities can give rise to heterogeneous nucleation.

(6) Results of zone refining

Later experiments showed that all the zone-refined materials were satisfactorily pure except salol in which gas bubbles were formed along the cell boundary during solidification. It is still uncertain, however, whether these gases came from the zone-refined material, or they were dissolved into the melt during the preparation of the specimen.

4.4 Preparation of the Specimen

4.4.1 The glass cell

The glass cell to be used to hold the specimen was made of two pieces of 24 x 24 x 0.15 (mm) microscope glass cover slips, which were cleaned with alcohol and dried before use. The glass slips were then stuck together with twinsticks. Two small openings were left at two

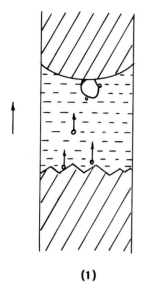 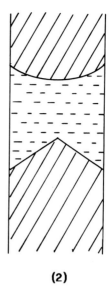

Fig. 4.14 (1) Early stage of zone refining.

(2) Later stage of zone refining.

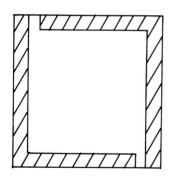 

Fig. 4.15 The glass cell.

Fig. 4.16 Filling the cell.

opposite corners: one for filling and the other for the escape of air, otherwise air would be entrapped. The glass cell (Fig. 4.15) was then dried.

4.4.2 The specimen

The material to be used was melted in a clean watchglass above a hot-plate. Then by dipping the opening at the corner of the glass cell into the melt, the melt will get into the cell by surface tension (Fig. 4.16). The cell was held close to the watchglass in order to keep warm to prevent the melt from freezing halfway. Care must be taken to ensure that the cell was filled completely with the melt. The specimen was cooled in the air. It was observed that, with the purified material, the melt was kept in the cell for several days without freezing; this was thought to be the result of high purity and thus lack of nuclei. Eventually a solid seed was fed at the opening and the material froze radiatively from the seed into the melt and grew rapidly. On the other hand, in the as-received materials, impurities exist, and the melt freeze itself from a number of sites and grow into several grains (Fig. 4.17). The outer surface of the cell was cleaned and dried; otherwise material residues could evaporate and obstruct illumination during experiment.

Other specimens (Fig. 4.18) were made by filling the melt into a capillary tube and inserting the tube into a glass cell, then filling the glass cell with the melt. Advantage was taken of the fact that better image could result from a glass tube containing and surrounded by melt. These specimens were used for faceted growth as well as for three dimensional non-faceted cellular array growth.

4.5 Optical considerations

The growth process was observed in bright field transmission (except otherwise specified) with a Wild M20 optical microscope; and photographed with an Exa-1a 35mm camera using Ilford 400 ASA film, usually exposed for 1/125 second. Objectives of magnification x3, x4 and x10, and a long working distance objective x32 are available. Eyepieces available included a x10 eyepiece and another x10 eyepiece with graticule. The magnification was calibrated by photographing a stage micrometer.

4.6 The experimental procedure

(i)   Material purification and specimen preparation.
(ii) Width of the gap between the hot-plate and the cold-plate was set. Specimen was put in place and capped.
(iii) Heating and water cooling started; microscope lighting on. The temperature gradient was calibrated for the first run.
(iii) About 5 hours for the temperature field to stabilize; input to heater then reached constant. Specimen first melted towards the hot-plate, and eventually a planar interface was established, which was above the gap between the hot plate and the cold plate.
(iv) Specimen was pulled at a set velocity towards the cold-plate; growth started.
(vii) Observation and photography.

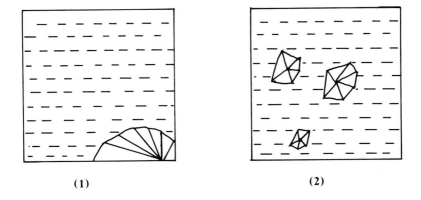

Fig. 4.17 Freezing of the sample.

(1) Purified material, and (2) raw material.

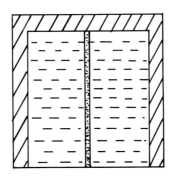

Fig. 4.18 Specimen contained in a tube sandwiched in the glass cell.

4.7 Summary

Most materials used in the work can be purified satisfactorily using the zone-refiner constructed. The temperature gradient stage can produce stable and adjustable growth conditions for *in situ* observation of the growth process. Growth parameters, *i.e.*, the temperature gradient and the moving velocity, have been calibrated.

Chapter V

EXPERIMENTAL RESULTS

5.1 Introduction

Using techniques and apparatus described in the previous chapter, both faceted and non-faceted cellular array growth were studied experimentally. The first group of experiments were conducted for faceting materials, including salol, thymol, and o-terphenyl. In these experiments, faceted cellular growth was produced under a number of growth conditions; the growth process was observed *in situ*; the growth mechanism was then studied. The results of these experiments revealed the important dynamical features of pattern formation in cellular growth. The second group of experiments were preliminary observations of non-faceted cellular array growth. These experiments were designed to be more accurately representative of steady state non-faceted cellular array growth as treated in typical analytical/numerical models.

5.2 Faceted Cellular Growth

5.2.1 Introduction

The problem of faceted cellular growth has just recently begun to receive considerable attention, in contrast to non-faceted cellular growth which has remained a subject of intensive research over the last few decades. The importance of understanding faceted cellular growth however cannot be overemphasized. Practically it represents an important aspect of the crystal growth of several important semiconductor materials, *i.e.*, Si, Ge, *etc*. On the fundamental side of the problem it represents a whole class of natural pattern formation in solidification, which is different from non-faceted cellular growth most commonly encountered in metals. Dynamic study of cellular growth as a whole has also recently aroused great interest. This is because the pattern selection problem cannot be well understood until vigorous dynamical study of the process of pattern development is carried out. The objective of the experimental work to be presented in this chapter was to observe faceted cellular growth *in situ* and to study the physical process and growth mechanisms, which provided the basis for a later theoretical study of the problem (see Chapters 8 & 9).

5.2.2 Results

5.2.2.1 Development of faceted cellular arrays (Figs. 5.1--5.3)

At the beginning of each experiment, the system was kept stationary in the temperature field for equilibration; eventually a planar solid/liquid interface was formed. As growth started, instabilities occurred on the originally planar interface, leading to its breakdown. Usually these instabilities first occurred at the grain or subgrain boundaries. As growth proceeded, these instabilities developed into cells bounded by facets on the sides and non-facets on the top. As growth continued, the top non-facet was eventually grown out, leading to the formation of a faceted cellular array. This occurred because facets have lower growth rates. The faceted cells

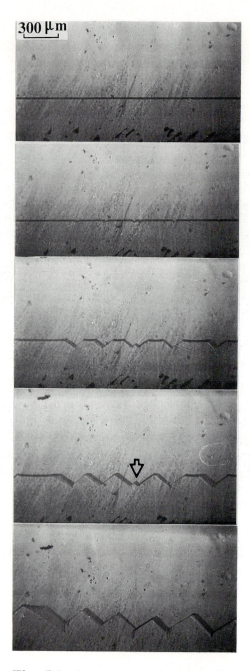

Fig. 5.1 A sequence showing the development of a faceted cellular array. (Thymol. $V = 1.1$ μm/s, $G = 52$ K/cm).

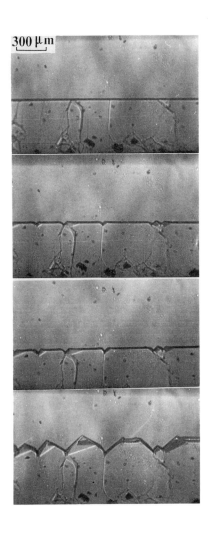
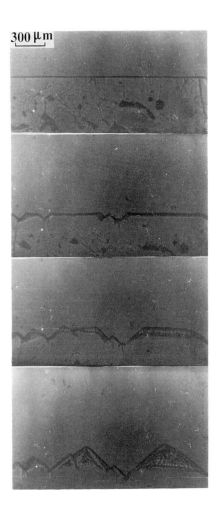

Fig. 5.2 A sequence showing the development of a faceted cellular array. (Thymol. V = 8.6 μm/s , G = 35 K/cm).

Fig. 5.3 A sequence showing the development of a faceted cellular array. (Thymol. V = 15.1 μm/s , G = 35 K/cm).

formed were either shallow cells (Fig. 5.1), or deep cells with deep liquid grooves along the cell boundaries (Figs. 5.2 & 5.3). The larger the growth rate, the severer the solute pile-up, and the deeper the groove.

It can also be seen that, during this process, some initially formed small cells were later grown out (as indicated by the arrow in Fig. 5.1). This means that, these small cells, though created by perturbations, were unable to survive. They lost to their larger neighbour cells during the growth competition, which started from the very beginning of the growth process.

5.2.2.2 Solute redistribution during growth

During growth solute was continuously rejected into the melt (if $k < 1$), leading to, besides the formation of liquid grooves, the formation of gas bubbles, especially in some cases with salol. Fig. 5.4 shows gas bubbles at the interface. At higher growth rates a gas bubble formed at the interface was entrapped at the cell boundary forming an elongated blowhole (Fig. 5.5). Round bubbles (Fig. 5.6) were formed if they were entrapped more rapidly. Fig. 5.6 gives an example of the distribution of entrapped gas bubbles. It can be seen from Fig. 5.7 that liquid later froze inside the gas bubble. This may be because liquid later migrated into the bubble and froze there. Fig. 5.8 shows that as the growth rate was increased the density of gas bubbles became higher.

It is thought that the relatively large contraction on freezing of the faceting organic compounds gives rise to the formation of gas bubbles, since it will locally reduce the pressure in the melt. In Fig. 5.9 cracks can be seen on the crystal, which was believed to result from the stress produced by the contraction on freezing.

A eutectic-like reaction was observed (Fig. 5.10), which is:

$$\text{Liquid} \text{ ------ } \text{Gas} + \text{'Eutectic'}$$

It can be seen that, the rod or fiber-like eutectic grew perpendicularly to the faceted interface. The 'rod' however does not follow a straight line; instead it follows a fishbone type path, similar in appearance to the fishbone type network found in silicon (Fig. 2.13). At the end of the 'rod' lies a gas bubble, which was observed under microscope to be jumping around along the facet as growth proceeded. It is thought that this has led to the fishbone path of the 'rod'. The average spacing of the 'rod' phase is almost constant, as can be seen from Fig. 5.10. This can be explained by the growth mechanism of the facet. The facet grows by the spreading of steps produced by emergent screw dislocations intersecting the facet. At the corner of the step solute is more concentrated, leading to the formation of the gas bubble (Fig. 5.11). As the step spreads over the facet, the bubble moves with it, leading to what was observed the jumping of bubbles along the facet. The 'rod' phase will also move with the step, and two 'rods' join when they meet. New bubbles and new 'rod' phase form at other steps. Since the spacing between steps should not be very different, the average spacing between the 'rod' phase appears approximately constant.

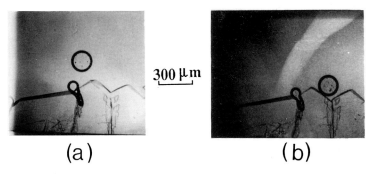

Fig. 5.4 Showing gas bubbles in front of the interface. Time interval approximately 10 seconds. (Salol. V = 8.6 µm/s , G = 18 K/cm).

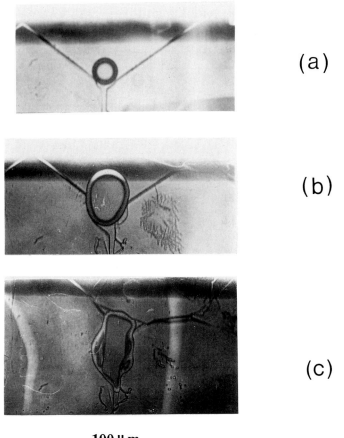

Fig. 5.5 A gas bubble at the cell boundary. Time interval between (a) and (b) approximately 3 seconds. (Salol. V = 15.1 µm/s , G = 18 K/cm).

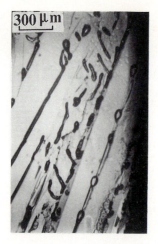 

Fig. 5.6 Round gas bubbles and elongated gas bubbles. (Salol. V = 28.3 μm/s , G = 18 K/cm).

Fig. 5.7 An entrapped gas bubble. (Salol. V = 28.3 μm/s , G = 18 K/cm).

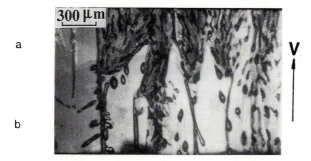

Fig. 5.8 Variation of the gas bubble distribution as the growth rate was increased. (Salol. G = 18 K/cm). (a) V = 28.3 μm/s , and (b) V = 8.6 μm/s .

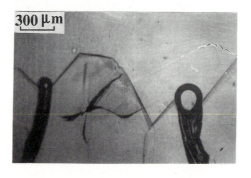

Fig. 5.9 Hot tear . (Salol)

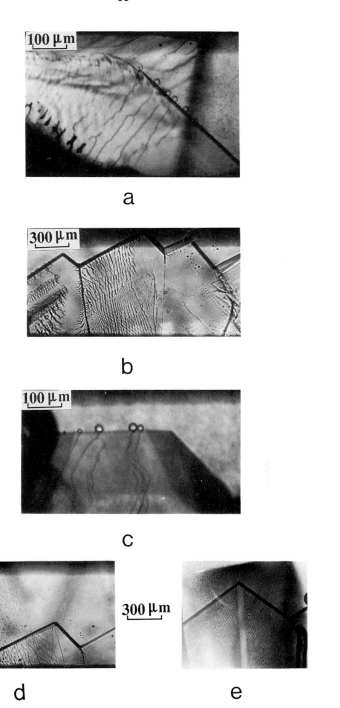

Fig. 5.10 Eutectic cells. (Salol). (a) V = 15.1 μm/s , G = 18 K/cm;
(b) V = 8.6 μm/s , G = 35 K/cm; (c) V = 15.1 μm/s , G = 18 K/cm;
(d) V = 8.6 μm/s , G = 65 K/cm; (e) V = 15.1 μm/s , G = 18 K/cm.

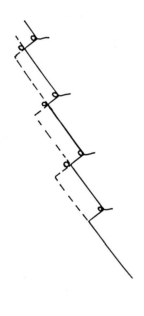

Fig. 5.11 Schematic diagram showing the growth of eutectic cells.

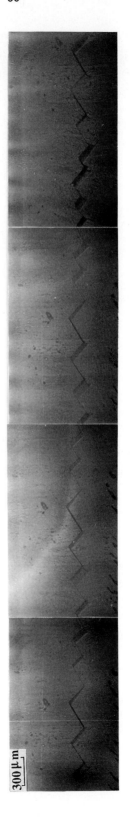

Fig. 5.12 A faceted cellular interface. (Thymol. V = 3.3 μm/s, G = 52 K/cm)

5.2.2.3 Morphology of the interface

Figs. 5.12-5.16 show faceted cellular interfaces developed under different growth conditions. It can be seen that at lower growth rates shallow cells tend to form (Fig. 5.12), while deep cells tend to form at higher growth rates. Fig. 5.18 shows the magnified view of a cell. The detailed structure of the facet can be seen. Fig. 5.17 shows the growth of the interface for o-terphenyl. As can be seen, the cellular interface is much more irregular, as this material is more faceting.

5.2.2.4 Growth competition between facets

Fig. 5.19 shows a faceted cell grown in a capillary tube. It demonstrates the swinging movement of the two facets forming the cell. The same movement has also been observed for a single crystal with a faceted interface in the zone refining sample. This movement results from the different growth rates of the two facets, leading to their competition. For cells bounded by facets, i.e. faceted cells, the variation of facet growth rates will inevitably lead to the interaction between cells in an array.

5.2.2.5 Cell interactions: formation of new cells

During growth it has been observed that both the number of cells and the size of an individual cell vary, i.e., cells interact; the interaction is more dramatic during a transient period, and more dramatic under some growth conditions than others although this cannot be clearly defined. The consequence of this interaction is the creation of new cells and loss of existing cells. It was observed that new cells can be created by splitting at a cell tip (Figs. 5.20-5.23). At first the tip of a cell was flattened and became non-faceted. This is because the cell tip cannot exceed the isotherm for $\Delta T_k = 0$ (i.e. the melting isotherm), where ΔT_k is the interface kinetic undercooling, taking the solute effect into account if any. The flattened cell tip will be further dumped with solute, and became unstable. New cells were consequently developed.

New cells can also be created through a 'side-branching' type behaviour of the faceted cell (Fig. 5.24), although this was observed less frequently than tip splitting. It occurs when a facet itself becomes unstable because of perturbations by foreign particles, or simply any perturbations the facet can be subject to. This 'side-branching' behaviour will lead to the formation of new cells. This occurs less frequently because facets are more stable than non-facets because of the kinetic effect [21,88,89]. It has also been observed that new cells can be created through nucleation at the base of the two intersecting facets for shallow cells.

5.2.2.6 Cell interactions: loss of cells

It was observed that sometimes a cell was overgrown by its neighbours in the array (Figs. 5.25-5.27). Fig. 5.25(b) shows the trace of the cell boundaries formed after two neighbouring cells had overgrown a smaller cell leading to the joining of the cell boundaries. Figs. 5.26 & 5.27 show sequences of the loss of cells. It can be seen that, for some reason a smaller cell became even smaller and eventually grown out by its neighbouring cells. Fig. 5.28 however

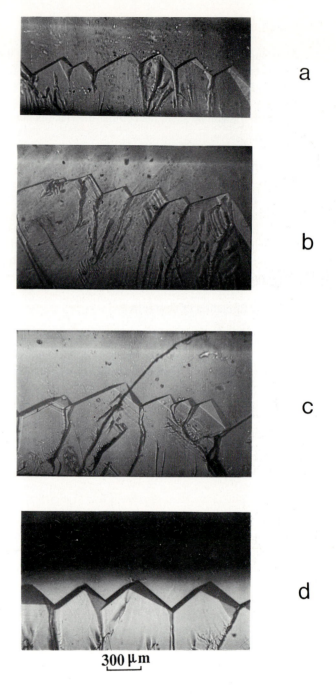

Fig. 5.13 Faceted cellular interfaces. (Thymol). (a) V = 5.7 µm/s, G = 35 K/cm; (b) V = 1.1 µm/s, G = 52 K/cm; (c) V = 6.4 µm/s, G = 52 K/cm; (d) V = 3.3 µm/s, G = 35 K/cm.

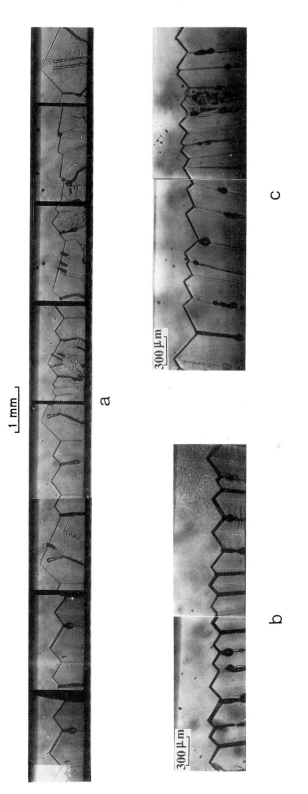

Fig. 5.14 Faceted cellular interfaces. (Salol). (a) V = 15.1 μm/s, G = 11 K/cm; (b) V = 8.6 μm/s, G = 18 K/cm; (c) V = 28.3 μm/s, G = 11 K/cm.

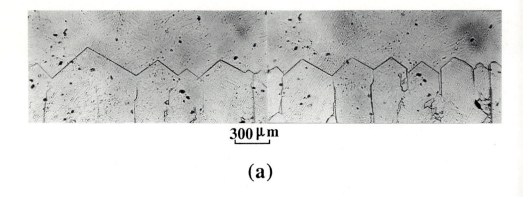

(a)

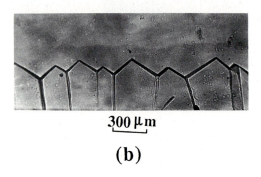

(b)

Fig. 5.15 Faceted cellular interfaces. (Salol). (a) V = 8.6 μm/s , G = 52 K/cm; (b) V = 28.3 μm/s , G = 18 K/cm.

Fig. 5.16 A faceted cellular interface. (Salol. V = 28.3 μm/s , G = 18 K/cm).

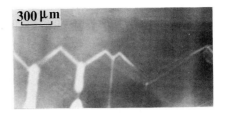

Fig. 5.17 A sequence showing faceted cellular growth in *o*-Terphenyl. (V = 3.3 μm/s , G = 35 K/cm).

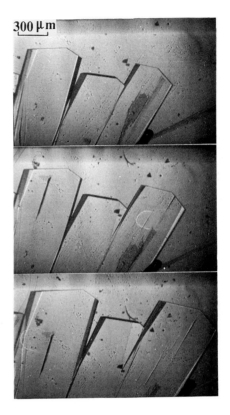

Fig. 5.18 Showing the detailed morphology of a cell tip. (Thymol)

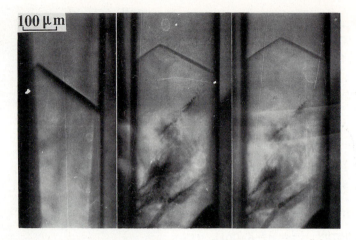

Fig. 5.19 A sequence of the growth of a faceted cell in a capillary tube showing the interaction between the two facets. (Salol. V = 8.6 μm/s , G = 18 K/cm).

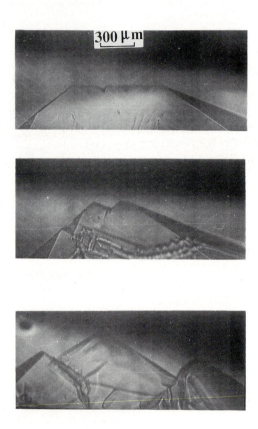

Fig. 5.20 A sequence showing the tip splitting of a cell. (Thymol. V = 3.3 μm/s , G = 35 K/cm).

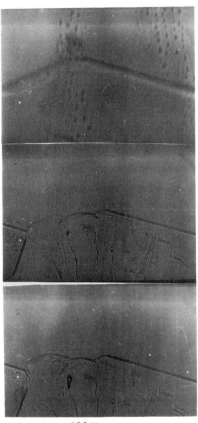

100 μm

Fig. 5.21 A sequence showing the tip splitting of a cell. (Salol. V = 28.3 μm/s , G = 18 K/cm).

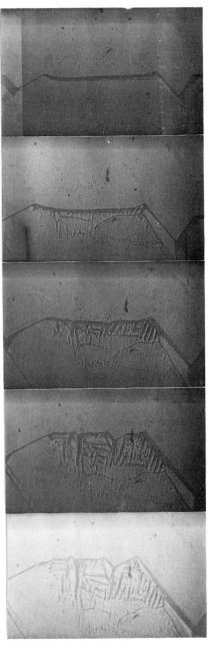

300 μm

Fig. 5.22 A sequence showing the tip splitting of a cell. (Thymol. V = 1.1 μm/s , G = 52 K/cm).

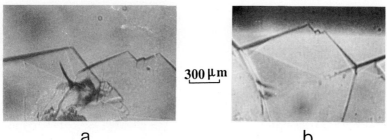

Fig. 5.23 Tip splitting of cells. (Salol)

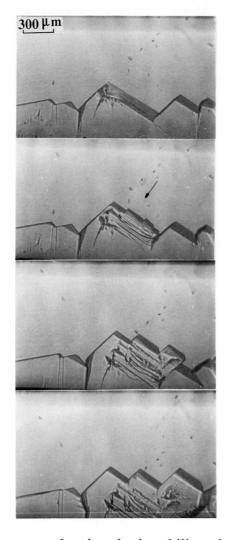

Fig. 5.24 A sequence showing the instability of a facet leading to the formation of new cells. (Thymol. V = 5.7 µm/s , G = 35 K/cm).

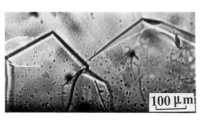 

Fig. 5.25 (a) Loss of a cell. (b) The trace of gas bubbles at the cell boundary after the loss of a cell. (Salol)

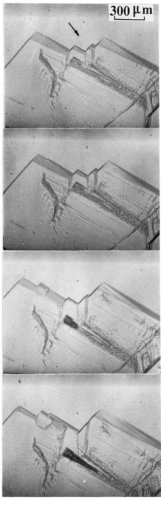

Fig. 5.26 A sequence showing the loss of cells in an array. (Thymol. V = 5.7 μm/s , G = 35 K/cm).

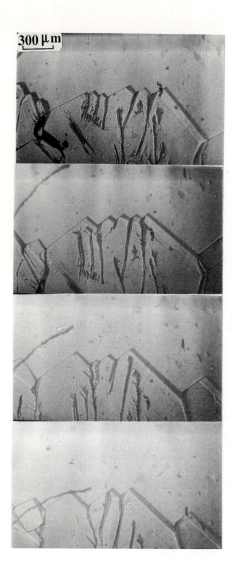

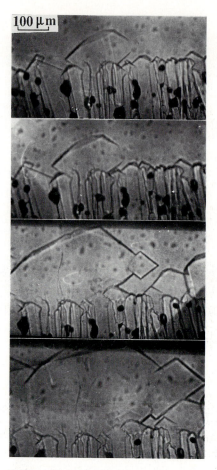

Fig. 5.28 Catastrophic overgrowth of cells. (Salol. V = 8.6 µm/s, G = 18 K/cm).

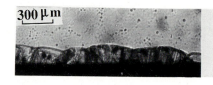

Fig. 5.27 A sequence showing the loss of cells in an array. (Thymol. V = 3.3 µm/s, G = 52 K/cm).

Fig. 5.29 A non-faceted interface formed at a high growth rate. (Salol. V = 60.2 µm/s, G = 11 K/cm).

illustrates that this can happen catastrophically if a cell somehow grows ahead of the others and gets a much larger growth rate.

5.2.2.7 Cell spacings

In the experiment, when growth reached steady state, the interface was photographed, from which the average cell spacing was later measured as the width of the interface, L, divided by the number of cells present, N (Fig. 5.30). Cells measured were mostly deep cells. Unfortunately there is considerable scatter making the measurement very difficult. It appears that there is a finite range of cell spacings under a given growth condition.

In Fig. 5.31, the average cell spacing thus measured was plotted against the growth rate under a number of temperature gradients. It is the middle value of the stable cell spacing range which was plotted in this figure (also listed in Table 5.1). The results indicate that the average cell spacing decreases with increasing the temperature gradient. Results with salol also indicate that the average cell spacing decreases with increasing the growth rate; results with thymol however are too few to give a clear indication.

When the growth rate is further increased, the interface was found to be non-faceted (Fig. 5.29). This suggests a faceting/non-faceting transition as the growth rate is increased to a certain extent.

5.2.3 Discussions and summary

The direct observation of faceted cellular growth has clearly revealed some important dynamical features of pattern formation. In a positive temperature gradient as is the case for directional solidification or zone melting, a planar solid/liquid interface becomes unstable because of constitutional undercooling. The relatively large kinetic undercooling term for faceted interface will however act as a stabilizing factor for the interface. An array of cells are formed after the planar interface breaks down. Shortly after the breakdown condition for a planar interface shallow cells are formed. Further away from the breakdown condition liquid grooves will form at the cell boundary (if $k < 1$) leading to the formation of deep cells. This is because, once solute can only be rejected sideways, the depth of the groove will be determined by a modified Scheil type expression, and is limited by a second phase being formed at the eutectic temperature or by a change in the distribution coefficient [48]. The groove however need not be faceted.

As soon as the array growth starts, cells in an array begin to interact in an attempt to approach the steady state under the specific growth condition. Depending on the current state of the system as compared with the steady state, an infinitesimal perturbation on the array can either be dampened out or amplified while being transmitted through the array. This is different from 'isolated' growth where such interactions do not exist.

The experimental results have indicated that the cell spacings spread over a certain range determined by the growth condition. This is because cellular interactions lead to the selection of cells, which is achieved through tip splitting for creation of new cells if the cell spacing is too

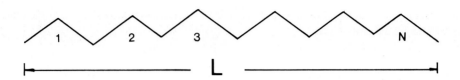

Fig. 5.30 Measurement of the average cell spacing.

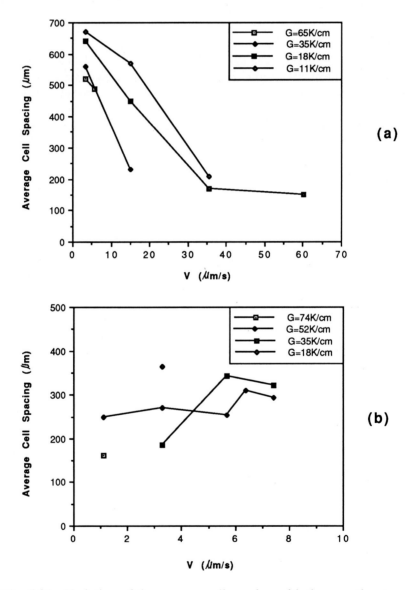

Fig. 5.31 Variation of the average cell spacing with the growth rate. (a) Salol (b) Thymol

Table 5.1   Average cell spacings

Material	V (µm/sec)	G (K/mm)	λ (µm)
Salol	3.3	6.5	520
	5.7	6.5	490
	3.3	3.5	560
	15.1	3.5	230
	3.3	1.8	640
	15.1	1.8	450
	35.4	1.8	170
	60.2	1.8	150
	3.3	1.1	670
	15.1	1.1	570
	35.4	1.1	210
Thymol	1.1	7.4	163
	1.1	5.2	248
	3.3	5.2	271
	5.7	5.2	254
	6.4	5.2	310
	7.4	5.2	294
	3.3	3.5	186
	5.7	3.5	343
	7.4	3.5	321
	3.3	1.8	365

large, or loss of cells if the cell spacing is too small. As a consequence only cells whose spacings lie within a certain range can survive.

The source for the interaction between cells is the variation in the facet growth rate. Cells with larger growth rates will increase their sizes at the expense of neighbouring cells whose growth rates are smaller. As a consequence the small cell will become smaller and smaller and eventually grown out. The increase in cell size, however, is limited by the fact that the cell tip cannot extend over the melting isotherm; *i.e.*, the cell tip will split if it becomes kinetically superheated. Moreover the facet itself will become unstable if it becomes too large. Cell splitting can also occur by perturbing the cell tip.

5.3 Preliminary Work with Non-Faceted Cellular Growth

5.3.1 Introduction

As can been seen from Chapter 2, there has been considerable amount of work on steady state non-faceted cellular array growth. The cellular shape has been calculated approximately by analytical models; more accurate numerical calculations have been reported more recently. Correspondingly a lot of experimental work has been carried out to compare with theoretical predictions, among which are thin film growth experiments with transparent organic compounds.

In this work non-faceted cells were grown in a capillary tube directionally (Fig. 4.18). It is thought that this should give a better representation of the 3-dimensional steady state cellular growth problem as treated by most of the mathematical models than the thin film growth experiment, as the latter actually represents a mixture of 2-D and 3-D growth. The cell shape obtained in this experiment should be comparable to theoretical predictions. Thin film experiments were carried out simultaneously outside the tube by growing the same material in the glass cell. Comparison of cellular growth can thus be made between the two regions.

5.3.2 Results and discussion

Succinonitrile was chosen for the experiment because it is the only transparent organic material available whose main thermodynamic properties have been reasonably accurately measured [57]. Experiments were carried out with zone-refined succinonitrile mixed up with 0.1% wt acetone. The temperature gradient was 65 K/cm. The capillary tube was 20 mm long and 167 µm in inner diameter.

At the beginning of each experiment, the specimen was equilibrated for more than 5 hours until a planar solid/liquid interface was formed. It was then moved at a certain velocity across the temperature field. Figs. 5.32--5.35 show sequences of the development of cellular interfaces at a number of growth rates. At $V = 1.1$ µm/sec, the planar front remained stable (Fig. 5.32). At $V = 3.9$ µm/sec, the planar front broke down and gradually developed into a cellular interface. A smooth cellular interface shape was formed as the steady state was reached (Figs. 5.33-5.34). As the growth velocity was further increased to 6.4 µm/sec, 'side-branching' instabilities occurred on the interface (Fig. 5.35).

Fig. 5.32 A sequence showing the development of the interface. (Succinonitrile-acetone. V = 1.1 μm/s, G = 65 K/cm). (a) 0; (b) 5; (c) 35; (d) 115 minutes.

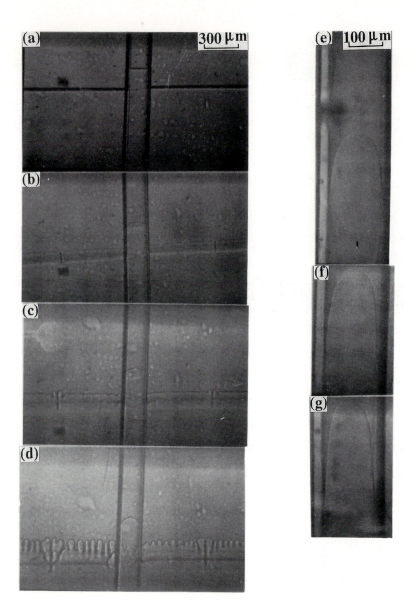

Fig. 5.33 A sequence showing the development of the interface. (Succinonitrile-acetone. V = 3.9 μm/s , G = 65 K/cm). (a) 0; (b) 3; (c) 4; (d) 5; (e) 9; (f) 15; (g) 20 minutes.

Fig. 5.34
A sequence showing the development of the interface. (Succinonitrile-acetone. V = 5.7 μm/s, G = 65 K/cm). (a) 0; (b) 1; (c) 6; (d) 21 minutes.

Fig. 5.35 A sequence showing the development of the interface. (Succinonitrile-acetone. V = 6.4 μm/s, G = 65 K/cm). (a) 16; (b) 21 minutes from beginning of growth.

Fig. 5.36 gives a comparison of the steady state cellular shapes formed at different growth rates. As can be seen, the steady state cellular interface became deeper and sharper as the growth rate was increased. This is to be expected since a larger growth rate requires a sharper cell tip to satisfy the diffusion requirement and a deeper groove to accommodate the rejected solute. However the sharpening of the cell tip will eventually be restrained by the surface energy term.

As can be seen from Fig. 5.37, cellular growth differed inside and outside the tube. Inside the tube the interface extended farther into the melt, *i.e.*, the interface undercooling was smaller, compared with outside the tube. The cell inside the tube was much sharper than the cells outside. A number of possible explanations for this difference were proposed: (1) Difference in heat transfer. However this was rejected by the observation that the melting fronts inside and outside the tube remained along the same line. (2) Difference in solute distribution caused by the different diffusion boundary conditions. However this is only true during the transient period. (3) The surface energy effect between succinonitrile and glass affects the interface shape differently for geometrical reasons. (4) They are both steady state solutions for cells growing under the same growth condition but at different undercoolings; however the different stability behaviours select different patterns: the cell inside the tube is 'isolated' and is not subject to long wave perturbations, while the cells outside are subject to both 'array stability' and 'tip stability'. (5) They are essentially different because the cell inside the tube is three dimensional while outside the tube the cells are two dimensional.

5.3.3 Summary

Preliminary results suggest that this experimental set-up should be more accurately representative of cellular array growth as treated in typical theoretical models, as it produces true three dimensional cells. It is suggested that caution must be exercised while comparison is made between theoretical predictions and thin film experiments.

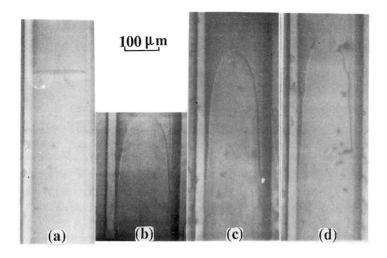

Fig. 5.36 Comparison of the steady state cellular shapes. (Succinonitrile-acetone. G = 65 K/cm). (a) V = 1.1 μm/s ; (b) V = 3.9 μm/s ; (c) V = 5.7 μm/s ; (d) V = 6.4 μm/s .

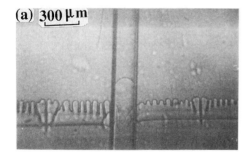 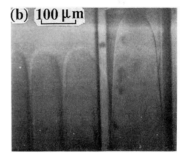

Fig. 5.37 Cellular interfaces inside and outside the tube. (Succinonitrile-acetone). (a) 5; (b) 20 minutes from beginning of growth. (V = 3.9 μm/s , G = 65 K/cm).

Chapter VI

NUMERICAL STUDY OF HEAT FLOW IN ZONE MELTING

6.1 Introduction

Zone melting (or zone refining) is a very important process of crystal growth. Heat flow is an important factor of the process, as it determines the shape of the molten zone which in turn affects the efficiency of zone refining purification; it also has strong influences on crystal growth, *e.g.*, the solid/liquid interface shape, stress, segregation, *etc*. The purpose of this study was to numerically model the heat flow in zone melting. Previous works [112-115] oversimplified the problem by neglecting the evolution/absorption of latent heat at the freezing/melting interfaces; however experimentally it has been found [116] that latent heat has a significant effect on the formation of the molten zone. In this work, heat flow in zone melting was modelled by an implicit method, while the release of latent heat was taken into account through an enthalpy method.

6.2 The Problem

The problem was schematically shown in Fig. 6.1. The material and its container, a glass tube, were heated uniformly over a distance, d, at the central region of the rod. The heater moves at a constant velocity V. The rod is long enough to assume that both ends well away from the heater is thermally insulated, *i.e.*, end effects are not included. Heat is transferred in the tube wall and the material inside the tube by conduction only with thermal conductivities independent of temperature (in fact temperature dependent thermal conductivities can be easily incorporated). Heat transfer between the tube wall and the surrounding air, and between the material and the tube wall, is assumed to obey the Newton's Law.

With sufficient heat input a molten zone will be produced in the neighbourhood of the heater. Two interfaces thus exist: a melting front absorbing latent heat and a freezing front releasing latent heat. Both interfaces are assumed to be along the melting isotherm, *i.e.*, no undercooling term is taken into account. This is a free moving boundary problem as the shapes of the two interfaces are not known a priori. In order to ease the difficulty in tracking the position of the interfaces, an enthalpy method can be used.

6.3 The Mathematical Formulation

6.3.1 The heat flow equation

Because of cylindrical symmetry only a slice of the rod needs to be treated (Fig. 6.2). The heat flow equation

$$\frac{\partial H}{\partial t} = k \nabla^2 T \qquad (6.1)$$

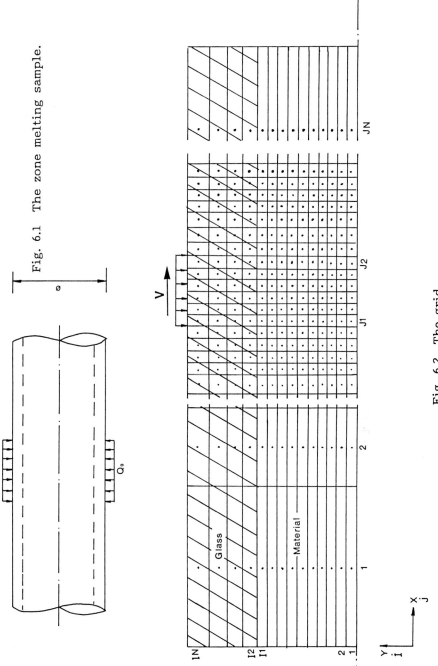

Fig. 6.1 The zone melting sample.

Fig. 6.2 The grid.

must be solved all over the whole domain under consideration taking the latent heat into account. This was done by a control volume finite difference method.

6.3.2 The discretization equations

As shown in Fig. 6.2, the whole region was divided into n boxes (n = IN x JN). Finer grids were used in the central region (*i.e.* near the heater) where phase changes will take place and heat flow is more intense. Farther away from the heater, heat flow is much less dramatic so larger boxes can be used to save computational cost. Mesh points were put halfway between box walls. Coordinates are employed which move with the heater at the velocity V.

At each mesh point (i, j), Eq. (6.1) was discretized as

$$\frac{\partial H}{\partial t} = Q = \sum_i Q_i \qquad i = E, W, N, S \qquad (6.2)$$

where E, W, N, and S denote the east, west, north, and south, respectively; Q_i is the heat flow across the wall. The flux term across each wall consists of a term due to the temperature gradient across the wall and a term due to movement of the wall for the east and west walls (Fig. 6.3), *i.e.*,

$$QE = \left[k \left(\frac{\partial T}{\partial x} \right)_E + V\, H_E \right] \Delta Y \qquad (6.3)$$

$$QW = \left[k \left(\frac{\partial T}{\partial x} \right)_W - V\, H_W \right] \Delta Y \qquad (6.4)$$

$$QN = k \left(\frac{\partial T}{\partial y} \right)_N \Delta X \qquad (6.5)$$

$$QS = k \left(\frac{\partial T}{\partial y} \right)_S \Delta X \qquad (6.6)$$

The temperature gradient across the wall and the temperature at the wall were obtained by linearly interpolating between the two adjacent points of that wall (Fig. 6.3). (An exponential scheme and a logarithmic scheme can be used for interpolations along the axial and radial directions respectively since they are the steady state solutions of the respective one dimensional heat conduction problems. This however was not presented here to avoid algebraic complexity in presentation.) Linear interpolation gives

$$\left(\frac{\partial T}{\partial x} \right)_E = \frac{T_{i, j+1} - T_{i, j}}{(\Delta X)_E} \qquad (6.7)$$

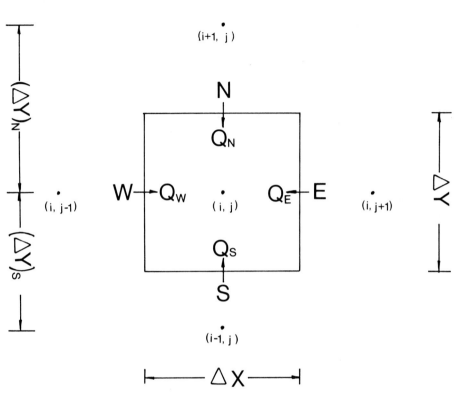

Fig. 6.3 Flux balance over a control volume.

$$\left(\frac{\partial T}{\partial x}\right)_W = \frac{T_{i, j-1} - T_{i, j}}{(\Delta X)_W} \tag{6.8}$$

$$\left(\frac{\partial T}{\partial y}\right)_N = \frac{T_{i+1, j} - T_{i, j}}{(\Delta Y)_N} \tag{6.9}$$

$$\left(\frac{\partial T}{\partial y}\right)_S = \frac{T_{i-1, j} - T_{i, j}}{(\Delta Y)_S} \tag{6.10}$$

$$T_E = T_{i, j} + \left(\frac{\partial T}{\partial x}\right)_E \frac{\Delta X}{2} \tag{6.11}$$

$$T_W = T_{i, j} + \left(\frac{\partial T}{\partial x}\right)_W \frac{\Delta X}{2} \tag{6.12}$$

H_E and H_W can be calculated from T_E and T_W, respectively (see 6.3.4) assuming

$$H_{(T = T_m)} = C T_m + L \tag{6.13}$$

Therefore, Q can be calculated from the temperature distribution. Note that different physical properties must be used for the glass region $(i \geq I_2)$ and the material region $(i \leq I_1)$.

6.3.3 Boundary conditions

The boundary conditions are schematically shown in Fig. 6.4.

(1) Centre of the rod (i = 1)

The symmetrical condition means that there is no heat flow across the south wall for boxes with i = 1, *i.e.*,

$$QS_{(i = 1)} = 0 \tag{6.14}$$

(2) The glass/material interface (i = I1, i = I2)

See Fig. 6.4 (2). T_1', T_2' represent the temperatures at the interface on the glass side and the material side respectively. It is assumed that heat flow between T_1' and T_2' obeys the Newton's Law, and there is no heat sink or source in the gap between them. Heat conservation thus requires

$$QN_2 = - QS_1 = QH = Q \tag{6.15}$$

where

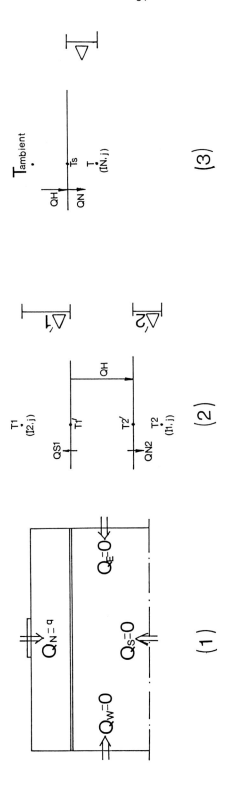

Fig. 6.4 Showing the boundary conditions.

$$QN_2 = k_2 \frac{T_2' - T_2}{\Delta_2} \tag{6.16}$$

$$QS_1 = k_1 \frac{T_1' - T_1}{\Delta_1} \tag{6.17}$$

$$QH = h_1 \left(T_1' - T_2'\right) \tag{6.18}$$

Solving these equations yields

$$Q = \frac{T_1 - T_2}{\frac{1}{h_1} + \frac{\Delta_2}{k_2} + \frac{\Delta_1}{k_1}} \tag{6.19}$$

Therefore

$$QN_{(i = I1)} = Q \tag{6.20}$$

$$QS_{(i = I2)} = -Q \tag{6.21}$$

(3) The surface (i = IN)

Similar arguments can be applied to the heat exchange between the outer surface and the surrounding air (Fig. 6.4 (3)), which is also assumed to obey the Newton's Law. Then we get

$$QN_i = QH \tag{6.22}$$

where

$$QN_i = k_1 \frac{T_S - T_i}{\Delta} \tag{6.23}$$

$$QH = -h_2 \left(T_S - T_{ambient}\right) \tag{6.24}$$

Solving these equations yields

$$QN_{(i = IN)} = \frac{T_{ambient} - T_i}{\frac{1}{h_2} + \frac{\Delta}{k_1}} \tag{6.25}$$

(4) The two ends (j = 1, j = JN)

These two ends are far away from the heater and are assumed to be thermally insulated. Therefore

$$QW_{(j=1)} = 0 \qquad (6.26)$$

$$QE_{(j=JN)} = 0 \qquad (6.27)$$

(5) Under the heater (i = IN and j = J1 to J2)

For the boxes under the heater, it is assumed that heat is supplied uniformly into these boxes through their north walls, *i.e.*,

$$QN_{(i=IN, j=J1 \text{ to } J2)} = q = \frac{Q_0}{d} \qquad (6.28)$$

6.3.4 The enthalpy method

In Fig. 6.5, enthalpy was plotted as a function of temperature for a pure material. Note that there is a discontinuity in the curve at the melting point T_m. The H (T) function can be explicitly expressed as:

For material (Fig. 6.5(1)):

$$H = \begin{cases} CT & T < T_m \\ CT+L & T > T_m \end{cases}$$

i.e.,

$$T = \begin{cases} H/C & H < CT_m \\ T_m & CT_m+L \geq H \geq CT_m \\ (H-L)/C & H > CT_m + L \end{cases} \qquad (6.29)$$

For glass (Fig. 6.5(2)):

$$H = C_g T$$

i.e.,

$$T = \frac{H}{C_g} \qquad (6.30)$$

as glass does not experience any phase change over the temperature range considered.

Given the heat content of a box, its temperature can be calculated from Eq. (6.29) or (6.30), assuming that the temperature at the grid point prevails throughout the whole box (which is the basic assumption of the finite difference method). The heat capacity C should be taken as that of the whole box, *i.e.*,

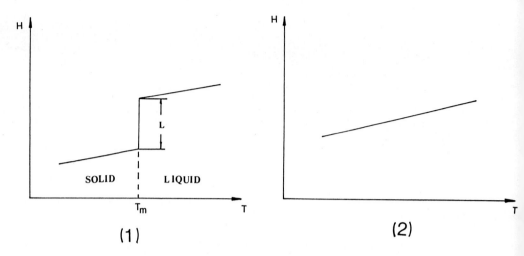

Fig. 6.5   The enthalpy function.   (1) Specimen   (2) Glass

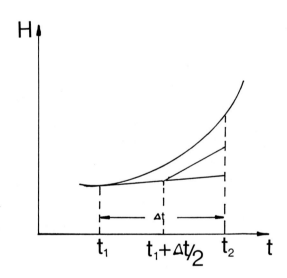

Fig. 6.6   The predictor-corrector method.

$$C = c\, \rho\, \Delta X\, \Delta Y \qquad (6.31)$$

At time t, from the current temperature distribution T, $\frac{\partial H}{\partial t}$ can be calculated from Eq. (6.2). If $\frac{\partial H}{\partial t}$ is constant over the whole time step Δt, the temperature distribution at the next time $t + \Delta t$ can then be calculated from this. The solution can thus be obtained by marching in time from a given initial temperature distribution. However, complexity arises because normally $\frac{\partial H}{\partial t}$ is not constant over a finite time step, especially when there is a phase change. This may be a source of error and thus of numerical instabilities, necessitating the formulation of an implicit scheme.

In the following sections the description is confined to the material region only for brevity; the glass region is much simpler since it doesn't undergo a phase change under the temperatures considered, and therefore can be treated in a similar but much simpler fashion.

6.3.5 Formulation of an implicit scheme

In order to calculate the heat content (or temperature) distribution at the next time step from the current temperature distribution through the heat flow equation, Eq. (6.2), some assumption must be made to calculate $\frac{\partial H}{\partial t}$. The explicit scheme assumes that $\frac{\partial H}{\partial t} = \left(\frac{\partial H}{\partial t}\right)_{t1}$, which is a first order approximation since it is only strictly valid if $\frac{\partial H}{\partial t} = $ constant. To ensure the stability of the explicit scheme small time steps and consequently large computational time are needed. It is conceivable that the discontinuity of the H-T curve at T_m will make the calculation more vulnerable to instability. The normal implicit scheme, on the other hand, assumes that $\frac{\partial H}{\partial t} = \left(\frac{\partial H}{\partial t}\right)_{t2}$. In this work, the following assumption is introduced:

$$\frac{\partial H}{\partial t} = \frac{1}{2}\left[(2 - \Omega)\left(\frac{\partial H}{\partial t}\right)_{t1} + \Omega \left(\frac{\partial H}{\partial t}\right)_{t2}\right] \qquad (6.32)$$

where Ω is an acceleration parameter $(\Omega < 2)$. For the sake of conciseness of presentation, we assume $\Omega = 1$, *i.e.*,

$$\frac{\partial H}{\partial t} = \frac{1}{2}\left[\left(\frac{\partial H}{\partial t}\right)_{t1} + \left(\frac{\partial H}{\partial t}\right)_{t2}\right] \qquad (6.33)$$

This is a second order approximation as it is strictly valid if $\frac{\partial^2 H}{\partial t^2}$ = constant; *i.e.*, the truncation error is of the order of Δt^3. This approximation is also known as a corrector-predictor method (Fig. 6.6). From Eq. (6.33) we have

$$\frac{H_{t2} - H_{t1}}{\Delta t} = \frac{1}{2}\left[\left(\frac{\partial H}{\partial t}\right)_{t1} + \left(\frac{\partial H}{\partial t}\right)_{t2}\right] \quad (6.34)$$

$$\left(\frac{\partial H}{\partial t}\right)_{t2} = 2\frac{H_{t2} - H_{t1}}{\Delta t} - \left(\frac{\partial H}{\partial t}\right)_{t1} \quad (6.35)$$

Substituting into Eq. (6.2) we get

$$2\frac{H_{t2} - H_{t1}}{\Delta t} - \left(\frac{\partial H}{\partial t}\right)_{t1} = Q_{t2} \quad (6.36)$$

$$2(H_{t2} - H_{t1}) - \left(\frac{\partial H}{\partial t}\right)_{t1}\Delta t - Q_{t2}\Delta t = 0 \quad (6.37)$$

Eq. (6.37) represents a set of nonlinear equations because of the non-linearity of H(T).

6.3.6 Solution of Eq. (6.37) by Newton-Raphson method

The set of nonlinear equations given by Eq. (6.37) can be solved by an iterative Newton-Raphson method [117]. The starting value for the iteration was taken as

$$H_{t2} = \left(\frac{\partial H}{\partial t}\right)_{t1}\Delta t \quad (6.38)$$

If this explicit prediction is right, it should satisfy Eq. (6.37). On the other hand, if it is wrong (as is normally the case), then it means that there is an error which leads to non-zero on the LHS of Eq. (6.37). The Newton-Raphson method can be used to remove this error iteratively. From Eq. (6.37) the error term for box k is

$$E_k = 2(H_{t2} - H_{t1})_k - \left(\frac{\partial H}{\partial t}\right)_{k,\,t1} - Q_{t2}\Delta t \quad (6.39)$$

From Eq. (6.39) we have

$$\frac{\partial E_k}{\partial H_p} = 2\,\delta(k, p) - \frac{\partial Q_k}{\partial T_p}\frac{\partial T_p}{\partial H_p}\Delta t \quad (6.40)$$

where $\delta(k, p)$ is the so-called Kronecker delta [117],

$$\delta(k, p) = \begin{cases} 0 & k \neq p \\ 1 & k = p \end{cases} \qquad (6.41)$$

From Eq. (6.30) we get

$$\frac{\partial T_p}{\partial H_p} = \begin{cases} 1/C & H < C\,T_m \\ 0 & C\,T_m \leq H \leq C\,T_m + L \\ 1/C & H > C\,T_m + L \end{cases} \qquad (6.42)$$

Noting that $E_k = f(H_p)$, $p = 1, 5$ representing the central point and the 4 immediate neighbour points of box k, we then have

$$\delta E_k = \sum_p \frac{\partial E_k}{\partial H_p} \delta H_p \qquad (6.43)$$

where δH_p is a correction term for H_p. In order to remove the error in Eq. (6.37) we must have

$$\delta E_k + E_k = 0 \qquad (6.44)$$

i.e.,

$$\sum_p \left[2\,\delta(k, p) - \frac{\partial Q_k}{\partial T_p} \frac{\partial T_p}{\partial H_p} \Delta t \right] \delta H_p + E_k = 0 \qquad (6.45)$$

Since $\delta(k, p)$, $\dfrac{\partial Q_k}{\partial T_p}$, and $\dfrac{\partial T_p}{\partial H_p}$ are all constants, Eq. (6.45) is a linear equation of δH_p. Applying this to all the points, k = 1, n, leads to a set of n linear equations of n unknowns, δH_p. This set of linear equations can then be solved by Gaussian elimination [117]. An improved value for H_k is thus obtained as

$$H_k = H_k\,(\text{old}) + \delta H_k \qquad (6.46)$$

This process is repeated until only a negligible error exists in Eq. (6.37). The solution is taken as H_{t2}, which can then be converted into T_{t2} through the enthalpy method, *i.e.* Eq. (6.29). The temperature field can thus be calculated by marching in time from a given initial temperature distribution.

6.4 The Computational Procedure

The computational procedure is shown schematically in Fig. 6.7. It consists of the following main steps:

(1) Initialization and discretization of the domain;

(2) Calculation of $\left(\frac{\partial H}{\partial t}\right)_{t1}$ from Eq. (6.2);

(3) Explicit prediction for H_{t2}: $H'_{t2} = \left(\frac{\partial H}{\partial t}\right)_{t1} \Delta t$;

(4) Setting up the Newton-Raphson equations for δH_k and solving the equations by Gaussian elimination;

(5) An improved value for H_{t2} as $H''_{t2} = H'_{t2} + \delta H_k$; and converting H_k into T_k through the enthalpy method (Eq. (6.29));

(6) Repeating (4) and (5) until δH_k becomes globally negligible; the solution is taken as H_{t2};

(7) Marching in time by repeating (2) to (6), until the steady state is achieved, i.e., $\frac{\partial H}{\partial t} = 0$

6.5 Results and Discussion

Numerical experiments were carried out, using IN x JN = 14 x 93, to evaluate the performance of the scheme. It was found that the scheme was stable except at very large time steps (about 100 times that allowed for an explicit scheme to ensure stability) when large oscillations occurred as the system just began to approach the steady state. This can be expected since the error introduced through Eq. (6.31) is of the order of $\Delta t^3 \frac{\partial^3 H}{\partial t^3}$, while $\frac{\partial^3 H}{\partial t^3}$ has its maximum value as T(t) just begins to approach the steady state. The acceleration parameter Ω can help with the speed of convergency if chosen properly.

The computer model developed was used to calculate the temperature distribution in the specimen as well as in the glass container, and to simulate the evolution of the molten zone, for a number of different experimental conditions (*e.g.*, Q_0, V, d, ϕ, k_1/k_2, *etc.*). Fig. 6.8 gives an example of the calculated steady state isotherms in the system (the isotherms were obtained through linear interpolation from the nodal values of the steady state temperature distribution). It can be seen that the size of the molten zone cannot be smaller than the radius of the rod. This is because heat is transferred at about the same rate in both directions. The calculated temperature distribution can be used in calculating the thermal stress distribution in the system.

It is suggested that the steady state solution of the transient problem should be the same as the solution to the corresponding steady state problem.

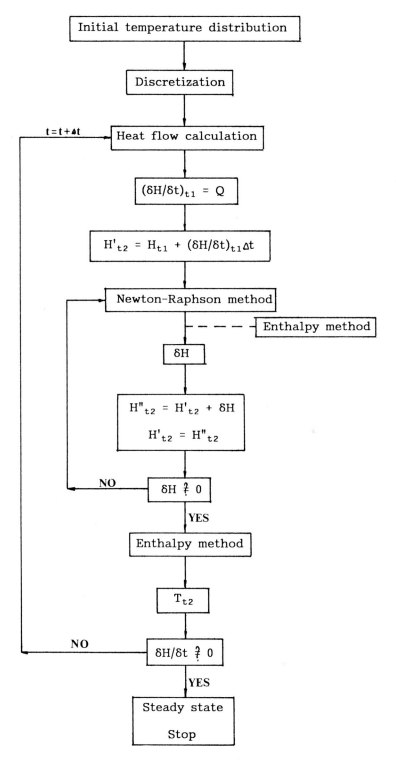

Fig. 6.7 The calculation procedure.

6.6 Summary

A numerical technique was developed to model the heat flow during zone melting with two co-existing moving solid/liquid interfaces, *i.e.* a melting interface and a freezing interface. In this technique, the latent heat evolution was included by an implicit finite difference scheme through the enthalpy method. The computer model developed was used to calculate the temperature distribution in the system and to follow the evolution of the molten zone for different experimental conditions.

Table 6.1 Symbol Table

Symbol	Meaning
c	Specific heat
C	Heat capacity
C_g	Heat capacity (glass)
d	Heater length
h_1	Heat transfer coefficient (glass/specimen)
h_2	Heat transfer coefficient (tube wall/air)
H	Heat content (Enthalpy)
k	Thermal conductivity
k_1	Thermal conductivity (glass)
k_2	Thermal conductivity (specimen)
L	Latent heat of fusion
Q	Heat flux
Q_0	Heat input
t	Time
T	Temperature
T_m	Melting temperature (specimen)
V	Heater velocity
X, Y	Coordinates
Ω	Acceleration parameter
ρ	Density

* Other symbols used are defined in the text.

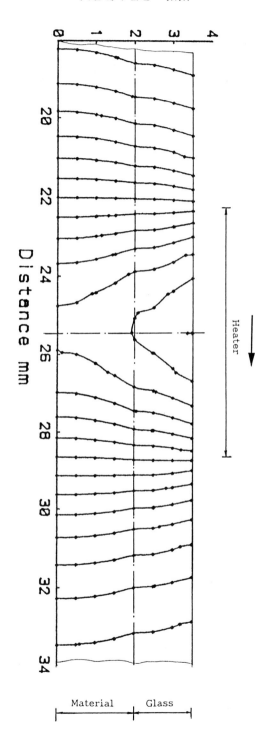

Fig. 6.8 The calculated isotherms.

Chapter VII

THEORETICAL STUDY OF STEADY STATE CELLULAR ARRAY GROWTH

7.1 Introduction

In a common approach, the problem of steady state cellular/dendritic growth consists mainly of three parts: (1) Solution of the solute diffusion field for an interface shape subject to appropriate boundary conditions; (2) Local coupling of the solute field with the imposed temperature field at the interface to obtain a self-consistent interface shape; and (3) Stability treatment, *i.e.*, application of the stability hypotheses, *e.g.* the extremum growth criterion or the marginal stability analysis, *etc.*, to determine the most stable solution which is supposed to exist in practice.

As was pointed out, array growth differs from isolated growth both in their steady state shapes and their stabilities, because the overlapping of the solute field in an array leads to the interaction between cells in an array. It is therefore essential to consider an array in order to describe cellular growth correctly. A number of approximate analytical models of cellular array growth have been set up (*e.g.* Refs. [40-42]). More recently numerical work has also been carried out by Hunt and McCartney [47,48] which calculates the self-consistent steady state three dimensional cell shape (see Chapter 2).

In the present work, an attempt was made to describe the solute diffusion for a given interface shape using the point source technique. In a collaborative effort, the analytical expressions were derived by Dr. H. Rauh, and evaluated numerically by the author. It is felt that this technique may be more accurate and could potentially be more rapid than the numerical techniques used in previous work. The characteristics of cellular array growth are represented by appropriate boundary conditions. This in fact represents a wide range of steady state solute/heat flow problems commonly encountered in solidification study. It is hoped that upon further development, the self-consistent cell shape can be calculated. In this chapter the treatment for two dimensional cells will be described. It is noted that the three dimensional problem will effectively become essentially two dimensional because of cylindrical symmetry.

7.2 The Problem

As shown in Fig. 7.1, a cell in an array is represented as a solid growing within a cylinder. A Cartesian coordinate system (X, Z) is introduced which is fixed at the symmetry boundary of the cell, and moves at the growth rate V. Because of radial symmetry only a half-strip of the cell needs to be treated. The interface shape is described by $Z = \varepsilon(X)$, where ε is a smooth or piecewise smooth function. Diffusion in the solid phase is neglected as it is much slower than that in the liquid phase. The steady state solute diffusion equation is:

$$\nabla^2 C + \frac{V}{D}\frac{\partial C}{\partial Z} = 0 \qquad (7.1)$$

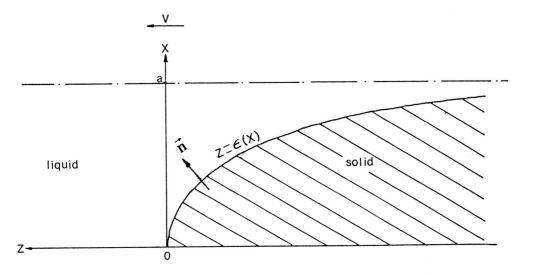

Fig. 7.1 A half-strip of the cell.

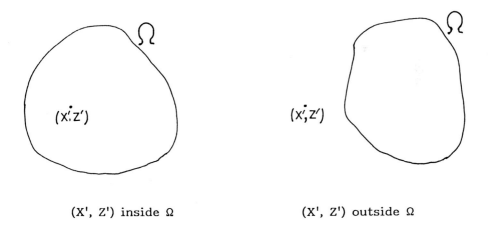

(X', Z') inside Ω   (X', Z') outside Ω

Fig. 7.2 An illustration of the meaning of the Green's function.

i.e.,

$$\frac{\partial^2 C}{\partial X^2} + \frac{\partial^2 C}{\partial Z^2} + \frac{V}{D}\frac{\partial C}{\partial Z} = 0 \tag{7.2}$$

with boundary conditions

$$\frac{\partial C}{\partial X} = 0 \quad \text{at } X = 0 \text{ and } X = a \tag{7.3}$$

Local equilibrium at the interface gives

$$C_{SI} = k\, C_I \tag{7.4}$$

Conservation of solute at the interface requires

$$V_n (1 - k)\, C_I = -D\frac{\partial C_I}{\partial n} \tag{7.5}$$

where the subscript 'I' denotes the interface.

Symmetry of the interface shape means

$$\frac{\partial \varepsilon}{\partial X} = 0 \quad \text{at } X = 0 \tag{7.6}$$

And finally, the far field condition is

$$C_{Z \to \infty} = C_\infty \tag{7.7}$$

7.3 The Technique

The technique of sources and sinks is used to solve the problem. This technique was originally developed for the problem of heat conduction [30]. The idea of the instantaneous point source of heat, that is, of a finite quantity of heat instantaneously liberated at a given point and time in an infinite solid, has proved most useful in the theory of conduction of heat. One great advantage is that it is based on a very simple physical idea, and this enables the solution of a large number of important problems to be written down immediately from first principles. Taking the solution for the instantaneous point source as fundamental, by integration w.r.t. time, we obtain the solution for the continuous point source, corresponding to release of heat at a given point at a prescribed rate. By integrating the solutions for point sources w.r.t. appropriate space variables, we obtain solutions for instantaneous and continuous line, plane, spherical surface sources, *etc*. By means of these solutions and the Green's theorem, the solution of the problem of conduction of heat for any given initial and boundary conditions, can be written down immediately in the form of definite integrals, taking the Green's function as the temperature at (X, Z) at time t due to an instantaneous point source of strength unity generated at the point (X', Z') at time t'. For steady state problems, however, the variable t will not appear. It is apparent that this technique should be equally applicable to the solute diffusion problem.

7.4 The Mathematical Formulation

7.4.1 The general scheme

The equation to be solved, Eq. (7.1), is a differential equation of order 2. The general scheme of the mathematical formulation is to use the Green's theorem to transform the equation into an integral equation while reducing the order of the differential equation. In order to apply the Green's theorem, it is necessary to introduce another function F, and to construct the Green's functions.

7.4.2 Normalization of variables

The following parameters and normalized variables are introduced to facilitate the formulation:

$$C^* = C - C_\infty \tag{7.8}$$

$$p = \frac{aV}{2D} \tag{7.9}$$

$$X^* = \frac{X}{a} \tag{7.10}$$

$$Z^* = \frac{Z}{a} \tag{7.11}$$

$$X'^* = \frac{X'}{a} \tag{7.12}$$

$$Z'^* = \frac{Z'}{a} \tag{7.13}$$

The following equations then result:

$$\frac{\partial^2 C^*}{\partial X^{*2}} + \frac{\partial^2 C^*}{\partial Z^{*2}} + 2p \frac{\partial C^*}{\partial Z^*} = 0 \tag{7.14}$$

$$\frac{\partial C^*}{\partial X^*} = 0 \quad \text{at } X^* = 0 \text{ and } X^* = 1 \tag{7.15}$$

$$C^*_{Z^* \to \infty} = 0 \tag{7.16}$$

$$\frac{\partial C_I^*}{\partial n^*} = -2p\frac{V_n}{V}(1-k)C_I \qquad \text{for } Z^* = \varepsilon(X^*) \qquad (7.17)$$

7.4.3 Construction of the Green's function for the problem stated by Eqs. (7.14)-(7.16)

The Green's function $G(X^*, Z^*, X'^*, Z'^*)$ in this case represents the concentration at (X^*, Z^*) due to a point source of strength unity at (X'^*, Z'^*), subject to the boundary conditions given by Eqs. (7.15)-(7.16). Thus we have

$$\frac{\partial^2 G}{\partial X^{*2}} + \frac{\partial^2 G}{\partial Z^{*2}} + 2p\frac{\partial G}{\partial Z^*} = -\delta(X^*, X'^*)\delta(Z^*, Z'^*) \qquad (7.18)$$

$$\frac{\partial G}{\partial X^*} = 0 \qquad \text{at } X^* = 0 \text{ and } X^* = 1 \qquad (7.19)$$

$$G_{Z^* \to \infty} = 0 \qquad (7.20)$$

where

$$\delta(m, n) = \begin{cases} 0 & m \neq n \\ 1 & m = n \end{cases} \qquad (7.21)$$

which is the so-called Kronecker Delta [117].

The meaning of the Green's function can be appreciated by integrating Eq. (7.18) over a domain Ω (Fig. 7.2):

$$\int_\Omega \text{LHS} = \int_\Omega \text{RHS} = \begin{cases} 0 & (X'^*, Z'^*) \text{ outside } \Omega \\ -1 & (X'^*, Z'^*) \text{ inside } \Omega \end{cases} \qquad (7.22)$$

Eq. (7.22) simply states that the total flux term across Ω equals the total source term inside domain Ω. This must hold under a steady state.

7.4.4 Solution of $G(X^*, Z^*, X'^*, Z'^*)$

The Green's function can be solved through a series of integral transforms.

7.4.4.1 Application of the finite Fourier cosine transform

The finite Fourier cosine transform states:

$$F[f(x)] \equiv f_n = \int_0^1 f(x) \cos(\lambda_n x) \, dx \tag{7.23}$$

$$F^{-1}[f_n] \equiv f(x) = \sum_{n=0}^{\infty} r_n f_n \cos(\lambda_n x) \tag{7.24}$$

if $\frac{df}{dx} = 0$ at $x = 0$ and $x = 1$, where

$$r_0 = 1, \quad r_n = 2, \quad n = 1, 2, 3, \ldots$$
$$\lambda_n = n\pi, \quad n = 0, 1, 2, 3, \ldots \tag{7.25}$$

It can be deduced that

$$F\left[\frac{d^2 f}{dx^2}\right] = -\lambda_n^2 \, F[f(x)] \tag{7.26}$$

Applying Eq. (7.26) to Eq. (7.18) yields

LHS:

$$\int_0^1 \left[\frac{\partial^2 G}{\partial X*^2} + \frac{\partial^2 G}{\partial Z*^2} + 2p \frac{\partial G}{\partial Z*}\right] \cos(\lambda_n X*) \, dX* = -\lambda_n^2 G_n + \frac{d^2 G_n}{dZ*^2} + 2p \frac{dG_n}{dZ*} \tag{7.27}$$

where $G_n(Z*) = F_{X*}[G(X*, Z*)]$

RHS:

$$\int_0^1 -\delta(X*, X'*) \, \delta(Z*, Z'*) \cos(\lambda_n X*) \, dX* = -\delta(Z*, Z'*) \cos(\lambda_n X'*)$$

$$(0 \le X'* \le 1) \tag{7.28}$$

Thus

$$\frac{d^2 G_n}{dZ*^2} + 2p \frac{dG_n}{dZ*} - \lambda_n^2 G_n = -\cos(\lambda_n X'*) \, \delta(Z*, Z'*) \tag{7.29}$$

7.4.4.2 Application of the complex Fourier transform

The complex Fourier transform states:

$$\tau[f(z)] \equiv f_k = \int_{-\infty}^{\infty} f(z) \exp(i\,K\,z)\,dz \tag{7.30}$$

$$\tau^{-1}[f_k] \equiv f(z) = \frac{1}{2\pi} \int_{-\infty - i\alpha}^{\infty - i\alpha} f_k \exp(-i\,K\,z)\,dK \tag{7.31}$$

It can be deduced that

$$\tau\left[\frac{df}{dz}\right] = -i\,K\,\tau[f(z)] \tag{7.32}$$

$$\tau\left[\frac{d^2f}{dz^2}\right] = -K^2\,\tau[f(z)] \tag{7.33}$$

Applying Eq. (7.33) to Eq. (7.29) gives

LHS:

$$\int_{-\infty}^{\infty} \left(\frac{d^2G_n}{dZ^{*2}} + 2\,p\,\frac{dG_n}{dZ^*} - \lambda_n^2\,G_n\right) \exp(i\,K\,Z^*)\,dZ^* = -K^2\,G_{n,K} - 2\,p\,i\,K\,G_{n,K} - \lambda_n^2\,G_{n,K} \tag{7.34}$$

where $G_{n,K} = \tau[G_n(Z^*)]$

RHS:

$$\int_{-\infty}^{\infty} -\cos(\lambda_n\,X'^*)\,\delta(Z^*,\,Z'^*)\exp(i\,K\,Z^*)\,dZ^* = \cos(\lambda_n\,X'^*)\exp(i\,K\,Z'^*) \tag{7.35}$$

Thus

$$K^2\,G_{n,K} + 2\,p\,i\,K\,G_{n,K} + \lambda_n^2\,G_{n,K} = \cos(\lambda_n\,X'^*)\exp(i\,K\,Z'^*) \tag{7.36}$$

$$G_{n,K}(X'^*,\,Z'^*) = \cos(\lambda_n\,X'^*)\,\frac{\exp(i\,K\,Z'^*)}{K^2 + 2\,p\,i\,K + \lambda_n^2} \tag{7.37}$$

7.4.4.3 Solution of G(X*, Z*, X'*, Z'*)

Applying the reverse complex Fourier transform Eq. (7.31) to Eq. (7.37) with

$$f_K = G_{n,K} \tag{7.38}$$

yields

$$G_n(Z^*) = \frac{1}{2\pi} \int_{-\infty-i\alpha}^{\infty-i\alpha} G_{nK} \exp(-iKZ^*) dK = \frac{\cos(\lambda_n X'^*)}{2\pi} \int_{-\infty-i\alpha}^{\infty-i\alpha} \frac{1}{K^2 + 2piK + \lambda_n^2} dK \quad (7.39)$$

By means of the Jordan's Lemma, it can be deduced that

$$G_n(Z^*) = \frac{\cos(\lambda_n X'^*)}{2\sqrt{p^2 + \lambda_n^2}} \exp\left[-p(Z^* - Z'^*) - \sqrt{p^2 + \lambda_n^2} |Z^* - Z'^*|\right] \quad (7.40)$$

Applying the reverse finite Fourier cosine transform Eq. (7.24) to Eq. (7.40) gives

$$G(X^*, Z^*, X'^*, Z'^*,) = \sum_{n=0}^{\infty} \frac{\eta_n}{\sqrt{p^2 + \lambda_n^2}} \cos(\lambda_n X^*) \cos(\lambda_n X'^*) \exp\left[-p(Z^* - Z'^*) - \sqrt{p^2 + \lambda_n^2} |Z^* - Z'^*|\right] \quad (7.41)$$

where $\eta_0 = 1/2$, $\eta_n = 1$, $n = 1, 2, 3, \ldots$

7.4.5 Introduction of the function F(X*, Z*)

The following function is introduced:

$$F(X^*, Z^*) = C^*(X^*, Z^*) \exp(pZ^*) \quad (7.42)$$

From Eq. (7.14) we have

$$\cap F(X^*, Z^*) = 0 \quad (7.43)$$

where

$$\cap \equiv \frac{\partial}{\partial X^{*2}} + \frac{\partial}{\partial Z^{*2}} - p^2 \quad (7.44)$$

The boundary conditions are:

$$\frac{\partial F}{\partial X^*} = 0 \quad \text{at } X^* = 0, \ X^* = 1 \quad (7.45)$$

$$V_n(1-k) C_I = -D \frac{\partial}{\partial n^*}[\exp(-pZ^*) F] \quad \text{for } Z^* = \varepsilon(X^*) \quad (7.46)$$

$$F_{Z^* \to \infty} = 0 \tag{7.47}$$

Eq. (7.47) holds because $C^*(X^*, Z^*)$ decays as $\exp(-2pZ^*)$ when $Z^* \to \infty$.

7.4.6 Construction of the Green's function $G^*(X^*,Z^*,X'^*,Z'^*)$ for $F(X^*,Z^*)$

$$\cap G^*(X^*, Z^*, X'^*, Z'^*) = -\delta(X^*, X'^*)\delta(Z^*, Z'^*) \tag{7.48}$$

with

$$\frac{\partial G^*}{\partial X^*} = 0 \quad \text{at } X^* = 0, \ X^* = 1 \tag{7.49}$$

$$G^*_{Z^* \to \infty} = 0 \tag{7.50}$$

As is done in (7.4.3), it can be obtained that

$$G^*(X^*,Z^*,X'^*,Z'^*) = \sum_{n=0}^{\infty} \frac{\eta_n}{\sqrt{p^2+\lambda_n^2}} \cos(\lambda_n X^*)\cos(\lambda_n X'^*)\exp\left[-\sqrt{p^2+\lambda_n^2}\,|Z^*-Z'^*|\right] \tag{7.51}$$

It can be seen from Eqs. (7.41) and (7.51) that

$$G(X^*, Z^*, X'^*, Z'^*) = \exp[-p(Z^* - Z'^*)] G^*(X^*, Z^*, X'^*, Z'^*) \tag{7.52}$$

It can also be seen that

$$G^*(X^*, Z^*, X'^*, Z'^*) = G^*(X'^*, Z'^*, X^*, Z^*) \tag{7.53}$$

i.e., G^* is symmetric. Therefore we have

$$\frac{\partial G^*}{\partial X'^*} = 0 \quad \text{at } X'^* = 0, \ X'^* = 1 \tag{7.54}$$

$$G^*_{Z'^* \to \infty} = 0 \tag{7.55}$$

7.4.7 Application of the Green's theorem

The second form of the Green's theorem [117] states:

$$\int_F \left[\phi(\underline{r}) \nabla^2 \theta(\underline{r}) - \theta(\underline{r}) \nabla^2 \phi(\underline{r})\right] df = \int_{\delta F} \left[\phi(\underline{r}) \underline{\nabla}\theta(\underline{r}) - \theta(\underline{r}) \underline{\nabla}\phi(\underline{r})\right] d\underline{\lambda} \quad (7.56)$$

where $\underline{\nabla} = \dfrac{d}{dn}$ ('_' denotes a vector.)

Since $\cap \equiv \nabla^2 - p^2$, it can be deduced that

$$\int_F \left[\phi(\underline{r}) \cap \theta(\underline{r}) - \theta(\underline{r}) \cap \phi(\underline{r})\right] df = \int_{\delta F} \left[\phi(\underline{r}) \underline{\nabla}\theta(\underline{r}) - \theta(\underline{r}) \underline{\nabla}\phi(\underline{r})\right] d\underline{\lambda} \quad (7.57)$$

In Cartesian coordinates Eq. (7.57) can be expressed as

$$\int\int_F \left[\phi(X'^*, Z'^*) \cap \theta(X'^*, Z'^*) - \theta(X'^*, Z'^*) \cap \phi(X'^*, Z'^*)\right] dX'^* \, dZ'^*$$

$$= \int_{\delta F} \left[\phi(X'^*, Z'^*) \frac{\partial \theta(X'^*, Z'^*)}{\partial n'^*} - \theta(X'^*, Z'^*) \frac{\partial \phi(X'^*, Z'^*)}{\partial n'^*}\right] ds'^* \quad (7.58)$$

where F denotes the area of integration, and δF denotes the respective boundary on which lies the line element given by

$$(ds'^*)^2 = (dX'^*)^2 + (dZ'^*)^2$$

Applying Eq. (7.58) with

$$\phi(X'^*, Z'^*) \equiv G^*(X^*, Z^*, X'^*, Z'^*) \quad (7.59)$$

$$\theta(X'^*, Z'^*) \equiv F(X'^*, Z'^*) \quad (7.60)$$

and noting that

$$\cap F(X'^*, Z'^*) = 0 \quad (7.61)$$

$$\int\int_F F(X'^*, Z'^*) \, \delta(X^*, X'^*) \, \delta(Z^*, Z'^*) \, dX'^* \, dZ'^* = F(X^*, Z^*) \quad (7.62)$$

we then get

$$F(X^*, Z^*) = \int_{\delta F}\left[G^*(X^*,Z^*,X'^*,Z'^*)\frac{\partial F(X'^*,Z'^*)}{\partial n'^*} - F(X'^*,Z'^*)\frac{\partial G^*(X^*,Z^*,X'^*,Z'^*)}{\partial n'^*}\right]ds'^*$$
(7.63)

Applying the boundary conditions (see Fig. 7.3)

$$\frac{\partial F}{\partial n'^*} = 0, \quad \frac{\partial G^*}{\partial n'^*} = 0 \quad \text{at } X'^* = 0, X'^* = 1, \text{ and as } Z'^* \to \infty \quad (7.64)$$

we get

$$F(X'^*,Z'^*) = \int_{Z'^*=\varepsilon(X'^*)}\left[G^*(X^*,Z^*,X'^*,Z'^*)\frac{\partial F(X'^*,Z'^*)}{\partial n'^*} - F(X'^*,Z'^*)\frac{\partial G^*(X^*,Z^*,X'^*,Z'^*)}{\partial n'^*}\right]ds'^*$$
(7.65)

7.4.8 Reversion to $C^*(X^*, Z^*)$

From Eqs. (7.42), (7.52) and (7.65) we have

$$\exp(p\,Z^*)\,C^*(X^*, Z^*) = \int_0^1 \{G(X^*,Z^*,X'^*,Z'^*)\exp[p(Z^*-Z'^*)]\frac{\partial[\exp(pZ'^*)C^*(X'^*,Z'^*)]}{\partial n'^*} -$$

$$- C^*(X'^*,Z'^*)\exp[pZ'^*]\frac{\partial\{\exp[p(Z^*-Z'^*)]G(X^*,Z^*,X'^*,Z'^*)\}}{\partial n'^*}\}_{Z'^*=\varepsilon(X'^*)}\left(\frac{ds'^*}{dX'^*}\right)dX'^*$$
(7.66)

where

$$\frac{ds'^*}{dX'^*} = \sqrt{1+\left(\frac{d\varepsilon}{dX'^*}\right)^2}$$

Noting that

$$\exp(-p\,Z'^*)\frac{\partial[\exp(p\,Z'^*)]}{\partial n'^*} = -p\frac{dX'^*}{ds'^*} \quad (7.67)$$

$$\exp(p\,Z'^*)\frac{\partial[\exp(-p\,Z'^*)]}{\partial n'^*} = p\frac{dX'^*}{ds'^*} \quad (7.68)$$

we get

$C^*(X^*,Z^*)$

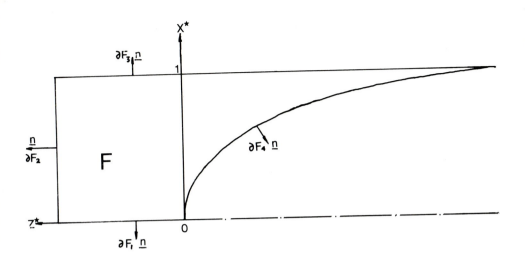

Fig. 7.3 The boundary condition.

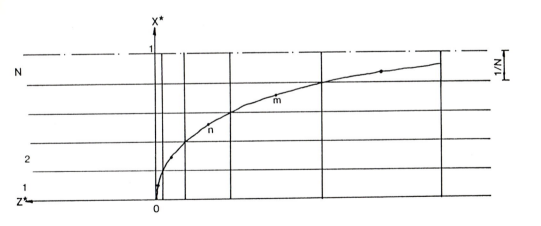

Fig. 7.5 The grid.

$$= \int_0^1 \left[G(X^*,Z^*,X'^*,Z'^*) \frac{\partial C^*(X'^*,Z'^*)}{\partial n'^*} - C^*(X'^*,Z'^*) \frac{\partial G(X^*,Z^*,X'^*,Z'^*)}{\partial n'^*} \right]_{Z'^*=\varepsilon(X'^*)} \left(\frac{ds'^*}{dX'^*}\right) dX'^*$$

$$- 2p \int_0^1 [G(X^*,Z^*,X'^*,Z'^*) C^*(X'^*,Z'^*)]_{Z'^*=\varepsilon(X'^*)} dX'^* \tag{7.69}$$

From Eq. (7.17) we get

$$C^*(X^*,Z^*) = 2p \int_0^1 \{G(X^*,Z^*,X'^*,Z'^*)[(1-k)C(X'^*,Z'^*) - C^*(X'^*,Z'^*)]\}_{Z'^*=\varepsilon(X'^*)} dX'^*$$

$$- \int_0^1 \left[C^*(X'^*, Z'^*) \frac{\partial G(X^*, Z^*, X'^*, Z'^*)}{\partial n'^*} \right]_{Z'^*=\varepsilon(X'^*)} \left(\frac{ds'^*}{dX'^*}\right) dX'^* \tag{7.70}$$

From Eq. (7.8) we have

$$C(X^*,Z^*) = H(X^*,Z^*) - 2pk \int_0^1 [G(X^*,Z^*,X'^*,Z'^*) C(X'^*,Z'^*)]_{Z'^*=\varepsilon(X'^*)} dX'^*$$

$$- \int_0^1 \left[C(X'^*, Z'^*) \frac{\partial G(X^*, Z^*, X'^*, Z'^*)}{\partial n'^*} \right]_{Z'^*=\varepsilon(X'^*)} \left(\frac{ds'^*}{dX'^*}\right) dX'^* \tag{7.71}$$

with

$$H(X^*, Z^*) = C_\infty + 2 p\, C_\infty \int_0^1 [G(X^*, Z^*, X'^*, Z'^*)]_{Z'^* = \varepsilon(X'^*)} dX'^*$$

$$+ C_\infty \int_0^1 \left[\frac{\partial G(X^*,Z^*,X'^*,Z'^*)}{\partial n'^*}\right]_{Z'^*=\varepsilon(X'^*)} \left(\frac{ds'^*}{dX'^*}\right) dX'^* \tag{7.72}$$

By introducing

$$C''(X^*, Z^*) = \frac{C(X^*, Z^*)}{C_\infty} \tag{7.73}$$

we then get

$$C''(X^*, Z^*) = H''(X^*, Z^*) - 2\,p\,k \int_0^1 [G(X^*, Z^*, X'^*, Z'^*)\, C''(X^*, Z^*)]_{Z'^* = \varepsilon(X'^*)}\, dX'^* -$$

$$- \int_0^1 \left[C''(X^*, Z^*) \frac{\partial G(X^*, Z^*, X'^*, Z'^*)}{\partial n'^*} \right]_{Z'^* = \varepsilon(X'^*)} \left(\frac{ds'^*}{dX'^*} \right) dX'^*$$

(7.74)

with

$$H''(X^*, Z^*) = 1 + 2\,p \int_0^1 [G(X^*, Z^*, X'^*, Z'^*)]_{Z'^* = \varepsilon(X'^*)}\, dX'^* +$$

$$+ \int_0^1 \left[\frac{\partial G(X^*, Z^*, X'^*, Z'^*)}{\partial n'^*} \right]_{Z'^* = \varepsilon(X'^*)} \left(\frac{ds'^*}{dX'^*} \right) dX'^*$$

(7.75)

7.5 Numerical Evaluation of Analytical Expressions

7.5.1 Evaluation of the Green's function and its derivatives

By taking the derivatives of Eq. (7.41), we get

$$\frac{\partial G}{\partial n'^*} = \left(\frac{ds'^*}{dX'^*} \right)^{-1} \left(\frac{\partial \varepsilon}{\partial X'^*}, -1 \right) \left(\frac{\partial G}{\partial X'^*}, \frac{\partial G}{\partial Z'^*} \right) = \left[\left(\frac{\partial \varepsilon}{\partial X'^*} \right) \left(\frac{\partial G}{\partial X'^*} \right) - \left(\frac{\partial G}{\partial Z'^*} \right) \right] \left(\frac{ds'^*}{dX'^*} \right)^{-1}$$

(7.76)

$$\frac{\partial G}{\partial X'^*} = \sum_{n=0}^{\infty} \frac{-\eta_n}{\sqrt{1+(p/\lambda_n)^2}} \cos(\lambda_n X^*) \sin(\lambda_n X'^*) \exp\!\left[-p(Z^*-Z'^*) - \sqrt{p^2+\lambda_n^2}\,|Z^*-Z'^*|\right]$$

(7.77)

$$\frac{\partial G}{\partial Z'^*} = \sum_{n=0}^{\infty} \eta_n \left[\frac{1}{\sqrt{1+(\lambda_n/p)^2}} + \mathrm{sgn}(Z^*-Z'^*) \right] \cos(\lambda_n X^*) \cos(\lambda_n X'^*) \cdot$$

$$\cdot \exp\!\left[-p(Z^*-Z'^*) - \sqrt{p^2+\lambda_n^2}\,|Z^*-Z'^*|\right]$$

(7.78)

The Green's function and its derivatives are made up of infinite series whose sum is unknown. Numerically they can only be evaluated by taking into account a finite number of terms. Fig. 7.4 shows plots of the Green's function and its derivatives evaluated numerically. It can be seen that, as the point of observation (X*, Z*) approaches the point of source (X'*, Z'*), singularities occur both in the Green's function and its derivatives. $G(X^*, Z^*, X'^*, Z'^*)$ has a logarithmic type singularity, while $\frac{\partial G}{\partial Z'^*}$ and $\frac{\partial G}{\partial X'^*}$ have 1/X type singularities. This is because the dampening effect of the exponential term in the series

$$\exp\left[-p(Z^* - Z'^*) - \sqrt{p^2 + \lambda_n^2}\,|Z^* - Z'^*|\right]$$

diminishes as $Z^* \to Z'^*$.

The Kummer's acceleration method [118] is used to accelerate the rate of convergence. This is done by introducing the following series whose sum is known in closed form:

$$R(X^*, Z^*, X'^*, Z'^*) = \sum_{n=1}^{\infty} \frac{1}{n\pi} \cos(n\pi X^*)\cos(n\pi X'^*)\exp(-n\pi|Z^* - Z'^*|)$$

$$= -\frac{1}{4\pi}\log\{1 - 2\exp(-\pi|Z^* - Z'^*|)\cos[\pi(X^* + X'^*)] + \exp(-2\pi|Z^* - Z'^*|)\} -$$

$$-\frac{1}{4\pi}\log\{1 - 2\exp(-\pi|Z^* - Z'^*|)\cos[\pi(X^* - X'^*)] + \exp(-2\pi|Z^* - Z'^*|)\}$$

(7.79)

Then we have

$$\frac{\partial R}{\partial X'^*} =$$

$$-\frac{1}{2}\exp(-\pi|Z^*-Z'^*|)\sin[\pi(X^*+X'^*)]\{1-2\exp(-\pi|Z^*-Z'^*|)\cos[\pi(X^*+X'^*)] + \exp(-2\pi|Z^*-Z'^*|)\}^{-1} +$$

$$+\frac{1}{2}\exp(-\pi|Z^*-Z'^*|)\sin[\pi(X^*-X'^*)]\{1-2\exp(-\pi|Z^*-Z'^*|)\cos[\pi(X^*-X'^*)] + \exp(-2\pi|Z^*-Z'^*|)\}^{-1}$$

(7.80)

$$\frac{\partial R}{\partial Z'^*} = \frac{1}{2}\,\text{sgn}(Z^*-Z'^*)\{\exp(-\pi|Z^*-Z'^*|)\cos[\pi(X^*+X'^*)] - \exp(-2\pi|Z^*-Z'^*|)\} \cdot$$

$$\cdot \{1 - 2\exp(-\pi|Z^* - Z'^*|)\cos[\pi(X^*+X'^*)] + \exp(-2\pi|Z^*-Z'^*|)\}^{-1} +$$

$$+\frac{1}{2}\,\text{sgn}(Z^*-Z'^*)\{\exp(-\pi|Z^*-Z'^*|)\cos[\pi(X^*-X'^*)] - \exp(-2\pi|Z^*-Z'^*|)\} \cdot$$

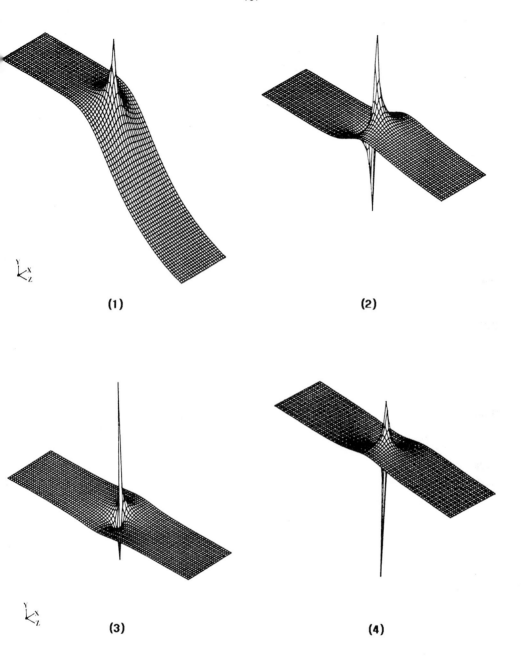

Fig. 7.4 The Green's function and its derivatives, for p = 0.5 and (X'*, Z'*) = (0.5, 0.0).

(1) $G(X^*, Z^*, X'^*, Z'^*)$ (2) $\dfrac{\partial G}{\partial Z'^*}$ (3) $\dfrac{\partial G}{\partial X'^*}$ (4) $\dfrac{\partial G}{\partial X'^*} + \dfrac{\partial G}{\partial Z'^*}$

$$\cdot \{1 - 2\exp(-\pi |Z^*-Z'^*|) \cos[\pi(X^*-X'^*)] + \exp(-2\pi |Z^*-Z'^*|)\}^{-1} \tag{7.81}$$

By introducing R into G, $\frac{\partial R}{\partial X'^*}$ into $\frac{\partial G}{\partial X'^*}$, and $\frac{\partial R}{\partial Z'^*}$ into $\frac{\partial G}{\partial Z'^*}$, respectively, after rearranging we get

$$G(X^*, Z^*, X'^*, Z'^*) = R(X^*, Z^*) + \frac{1}{2p} \exp[-p(Z^* - Z'^* + |Z^*- Z'^*|)] +$$

$$+ \sum_{n=1}^{\infty} \cos(\lambda_n X^*)\cos(\lambda_n X'^*) \left\{ \frac{\exp[-p(Z^*-Z'^*)-\sqrt{p^2+\lambda_n^2}\,|Z^*-Z'^*|]}{\sqrt{p^2+\lambda_n^2}} - \frac{\exp(-\lambda_n|Z^*-Z'^*|)}{\lambda_n} \right\} \tag{7.82}$$

$$\frac{\partial G}{\partial X'^*} = \frac{\partial R}{\partial X'^*} -$$

$$- \sum_{n=1}^{\infty} \cos(\lambda_n X^*)\sin(\lambda_n X'^*) \left\{ \frac{\lambda_n}{\sqrt{p^2+\lambda_n^2}} \exp[-p(Z^*-Z'^*)-\sqrt{p^2+\lambda_n^2}\,|Z^*-Z'^*|] - \exp(-\lambda_n|Z^*-Z'^*|) \right\} \tag{7.83}$$

$$\frac{\partial G}{\partial Z'^*} = \frac{\partial R}{\partial Z'^*} + \frac{1}{2}[1+\mathrm{sgn}(Z^*-Z'^*)]\exp[-p(Z^*-Z'^*+|Z^*-Z'^*|)] + \sum_{n=1}^{\infty} \cos(\lambda_n X^*)\cos(\lambda_n X'^*) \cdot$$

$$\cdot \left\{ \left[\frac{p}{\sqrt{p^2+\lambda_n^2}} + \mathrm{sgn}(Z^*-Z'^*)\right] \exp[-p(Z^*-Z'^*)-\sqrt{p^2+\lambda_n^2}\,|Z^*-Z'^*|] - \mathrm{sgn}(Z^*-Z'^*)\exp(-\lambda_n|Z^*-Z'^*|) \right\} \tag{7.84}$$

In Table 7.1 results of numerical evaluation of the Green's function are listed. As can be seen, the original form of the Green's function (without acceleration) converges very slowly as the point of observation gets nearer the point of source. With acceleration, however, the infinite series converges much more rapidly, especially around the singularities. This means that fewer terms are needed for numerically calculating the series for a given accuracy. It can also be seen that, with increasing the Peclet number p, the rate of convergence decreases.

7.5.2 Solution of the solute field

7.5.2.1 The integral equation

It can be seen from Eq. (7.74) that, given the boundary conditions, the solute field is fully determined by the interface concentration, *i.e.*, the source term. That is to say, once the interface concentration is obtained, the solute field can be calculated from Eq. (7.74). Applying Eqs. (7.74) & (7.75) at the interface $Z^* = \varepsilon(X^*)$ yields

Table 7.1 Results of Numerical Evaluation of Green's Function

(X*, Z*): Point of observation (X'*, Z'*): Point of source
p: Peclet number NMAX: Maximum order of modes taken into account
G: Value of Green's function G': Value of Green's function with acceleration

(X*,Z*)	(X'*,Z'*)	NMAX=1 (p=0.01)	NMAX=10 (p=0.01)
(0.0,-1.0)	(0.5,0.0)	G 50.000000000000 G' 49.999703064427	49.999700082943 49.999700082943
(0.5,-1.0)	(0.5,0.0)	G 50.000000000000 G' 50.000297490602	50.000300477661 50.000300477661
(0.95,-1.0)	(0.5,0.0)	G 50.000000000000 G' 49.999717558196	49.999714722241 49.999714722241
(0.0,-0.5)	(0.5,0.0)	G 50.000000000000 G' 49.993266750553	49.993233035751 49.993233035918
(0.5,-0.5)	(0.5,0.0)	G 50.000000000000 G' 50.007030740049	50.007065944992 50.007065945171
(0.95,-0.5)	(0.5,0.0)	G 50.000000000000 G' 49.993576661779	49.993544498919 49.993544498869
(0.0,0.0)	(0.5,0.0)	G 50.000000000000 G' 49.889682199924	49.875328810215 49.889682382228
(0.5,0.0)	(0.5,0.0)	G 50.000000000000	50.363403547731
(0.95,0.0)	(0.5,0.0)	G 50.000000000000 G' 49.891653823413	49.894126775271 49.891653998150
(0.0,0.5)	(0.5,0.0)	G 49.502491687458	49.495792055629
(0.5,0.5)	(0.5,0.0)	G 49.502491687458 G' 49.509522427507	49.509487325123 49.509487325302
(0.95,0.5)	(0.5,0.0)	G 49.502491687458 G' 49.496068349237	49.496100419687 49.496100419637
(0.0,1.0)	(0.5,0.0)	G 49.009933665338 G' 49.009636729765	49.009639687036 49.009639687036
(0.5,1.0)	(0.5,0.0)	G 49.009933665338 G' 49.010231155939	49.010228193143 49.010228193143
(0.95,1.0)	(0.5,0.0)	G 49.009933665338 G' 49.009651223534	49.009654036457 49.009654036457
(0.0,2.0)	(0.5,0.0)	G 48.039471957616 G' 48.039471402589	48.039471413589 48.039471413589
(0.5,2.0)	(0.5,0.0)	G 48.039471957616 G' 48.039472512645	48.039472501645 48.039472501645
(0.95,2.0)	(0.0,0.5)	G 48.039471957616 G' 48.039471429754	48.039471440215 48.039471440215

Table 7.1 (continued)

(X*,Z*)	(X'*,Z'*)		NMAX=1 (p=0.1)	NMAX=10 (p=0.1)
(0.0,-1.0)	(0.5,0.0)	G G'	5.000000000000 4.999703064427	4.999672138137 4.999672138137
(0.5,-1.0)	(0.5,0.0)	G G'	5.000000000000 5.000297490602	5.000328475001 5.000328475001
(0.95,-1.0)	(0.5,0.0)	G G'	5.000000000000 4.999717558196	4.999688141433 4.999688141433
(0.0,-0.5)	(0.5,0.0)	G G'	5.000000000000 4.993266750553	4.992925285219 4.992925285386
(0.5,-0.5)	(0.5,0.0)	G G'	5.000000000000 5.007030740049	5.007387385613 5.007387385792
(0.95,-0.5)	(0.5,0.0)	G G'	5.000000000000 4.993576661779	4.993250907219 4.993250907169
(0.0,0.0)	(0.5,0.0)	G G'	5.000000000000 4.889682199924	4.875346854612 4.889700426624
(0.5,0.0)	(0.5,0.0)	G	5.000000000000	5.363379891058
(0.95,0.0)	(0.5,0.0)	G G'	5.000000000000 4.891653823413	4.894144070678 4.891671293557
(0.0,0.5)	(0.5,0.0)	G	4.524187090180	4.517785623524
(0.5,0.5)	(0.5,0.0)	G G'	4.524187090180 4.531217830228	4.530871473104 4.530871473283
(0.95,0.5)	(0.5,0.0)	G G'	4.524187090180 4.517763751958	4.518080258494 4.518080258444
(0.0,1.0)	(0.5,0.0)	G G'	4.093653765390 4.093356829817	4.093385334800 4.093385334800
(0.5,1.0)	(0.5,0.0)	G G'	4.093653765390 4.093951255991	4.093922697975 4.093922697975
(0.95,1.0)	(0.5,0.0)	G G'	4.093653765390 4.093371323586	4.093398437190 4.093398437190
(0.0,2.0)	(0.5,0.0)	G G'	3.351600230178 3.351599675151	3.351599776541 3.351599776541
(0.5,2.0)	(0.5,0.0)	G G'	3.351600230178 3.351600785207	3.351600683817 3.351600683817
(0.95,2.0)	(0.5,0.0)	G G'	3.351600230178 3.351599702316	3.351599798743 3.351599798743

Table 7.1 (continued)

(X*,Z*)	(X'*,Z'*)		NMAX=1 (p=1.0)	NMAX=10 (p=1.0)
(0.0,-1.0)	(0.5,0.0)	G G'	0.500000000000 0.499703064427	0.499263520506 0.499263520506
(0.5,-1.0)	(0.5,0.0)	G G'	0.500000000000 0.500297490602	0.500737924889 0.500737924889
(0.95,-1.0)	(0.5,0.0)	G G'	0.500000000000 0.499717558196	0.499299464057 0.499299464057
(0.0,-0.5)	(0.5,0.0)	G G'	0.500000000000 0.493266750553	0.489468363731 0.489468363898
(0.5,-0.5)	(0.5,0.0)	G G'	0.500000000000 0.507030740049	0.511010958806 0.511010958985
(0.95,-0.5)	(0.5,0.0)	G G'	0.500000000000 0.493576661779	0.489952198485 0.489952198435
(0.0,0.0)	(0.5,0.0)	G G'	0.500000000000 0.389682199924	0.377115225165 0.391468797177
(0.5,0.0)	(0.5,0.0)	G	0.500000000000	0.861052723485
(0.95,0.0)	(0.5,0.0)	G G'	0.500000000000 0.391653823413	0.395839184622 0.393366407501
(0.0,0.5)	(0.5,0.0)	G	0.183939720586	0.180065348120
(0.5,0.5)	(0.5,0.0)	G G'	0.183939720586 0.190970460634	0.187990425958 0.187990426137
(0.95,0.5)	(0.5,0.0)	G G'	0.183939720586 0.177516382364	0.180243340979 0.180243340930
(0.0,1.0)	(0.5,0.0)	G G'	0.067667641618 0.067370706045	0.067567969957 0.067567969957
(0.5,1.0)	(0.5,0.0)	G G'	0.067667641618 0.067965132220	0.067767508892 0.067767508892
(0.95,1.0)	(0.5,0.0)	G G'	0.067667641618 0.067385199814	0.067572834388 0.067572834388
(0.0,2.0)	(0.5,0.0)	G G'	0.009157819444 0.009157264418	0.009157756115 0.009157756115
(0.5,2.0)	(0.5,0.0)	G G'	0.009157819444 0.009158374473	0.009157882774 0.009157882774
(0.95,2.0)	(0.5,0.0)	G G'	0.009157819444 0.009157291582	0.009157759215 0.009157759215

$$C_I^{''}(X^*) = H_I^{''} - \int_0^1 K_I(X^*, X'^*) C_I^{''}(X'^*) \, dX'^* \quad (0 \le X^* \le 1) \quad (7.85)$$

$$H_I^{''}(X^*) = 1 + \int_0^1 L_I(X^*, X'^*) \, dX'^* \quad (7.86)$$

with

$$L_I(X^*, X'^*) = 2 p \, G_I(X^*, X'^*) + \left(\frac{ds'^*}{dX'^*}\right)\left[\frac{\partial G_I(X^*, X'^*)}{\partial n'^*}\right] \quad (7.87)$$

$$K_I(X^*, X'^*) = 2 p k \, G_I(X^*, X'^*) + \left(\frac{ds'^*}{dX'^*}\right)\left[\frac{\partial G_I(X^*, X'^*)}{\partial n'^*}\right] \quad (7.88)$$

In Eq. (7.85) the interface concentration appears on both sides of the equation as the source term and the observation term. This is an integral equation and can only be solved numerically.

7.5.2.2 Discretization

In order to solve the integral equation Eq. (7.85) numerically, the equation must be discretized. As shown in Fig. 7.5, the interface is divided equidistantly into N sections with grid points halfway between walls.

7.5.2.3 Linearization of the equation

Assuming that the concentration $C_n^{''}$ at the grid point n prevails throughout the whole section, Eq. (7.85) can be rewritten as

$$C_m^{''} = H_{I\,m}^{''} - \sum_{n=1}^{N} K_{mn} C_n^{''} \quad m = 1, 2, 3, \ldots, N \quad (7.89)$$

where

$$K_{mn} = \int_{(n-1)/N}^{n/N} K_I(X_m^*, X'^*) \, dX'^* \quad (7.90)$$

with (see Fig. 7.5)

$$X_m^* = \frac{m-1}{N} + \frac{1}{2N} \quad (7.91)$$

By rearranging Eq. (7.89) we have

$$\sum_{n=1}^{N} A^*_{mn} C''_n = H''_{I\,m} \qquad m = 1, 2, 3, \ldots, N \quad (7.92)$$

with

$$A^*_{mn} = K_{mn} + \delta(m, n) \qquad (7.93)$$

7.5.2.4 Solution for the interface concentration

Eq. (7.92) represents a set of N linear equations of N unknowns C''_n. This set of linear equations can be solved by Gaussian elimination [117]. Once the interface concentration is obtained, the solute field can be calculated.

7.5.3 The computational effort

7.5.3.1 Numerical integration

For performing the numerical integration, proper care must be taken with the singularities in the integrand of the integration. As mentioned earlier these singularities occur when the point of observation approaches the point of source. The integration here is understood in the principal value sense. The integration interval was divided such that singularities only occur at the edge of the integration interval. Numerical integration was then performed using a NAG subroutine [119] D01AJF which is a general-purpose integrator to calculate an approximation to the integral of a function over a finite interval. This subroutine uses the Gauss 10-point and Kronrod 21-point rules, and can be used when the integrand has singularities at the edge of the integration interval, especially when these singularities are of algebraic or logarithmic type.

7.5.3.2 Solution of linear equations

The linear equations were solved using a NAG subroutine F04ATF which solves linear equations by a Crout's factorization method (a version of the Gaussian elimination method).

7.6 Results

In order to evaluate the performance of the technique, numerical experiments were performed for a shallow faceted cell (Figs. 7.6 & 7.7) and a deep cell with a asymptotic interface (Fig. 7.8(1)), *i.e.*,

$$Z^* = \log_e \left[\cos \left(\frac{\pi}{2} X^* \right) \right] \qquad (7.94)$$

under a number of experimental conditions. In these experiments the maximum order of modes taken into account in the series was 10. The interface was divided into 70 sections for the shallow cell and 150 for the deep cell. The accuracy specified for the NAG subroutine was

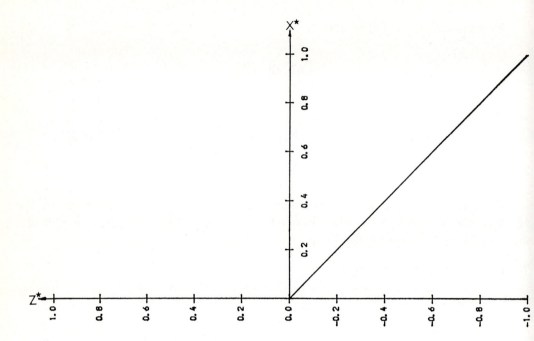

Fig. 7.6 (1) The flat interface.

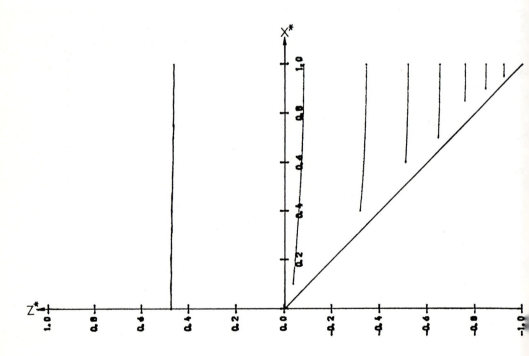

Fig. 7.6 (2) Isoconcentrates (p = 1.0, k = 0.1).

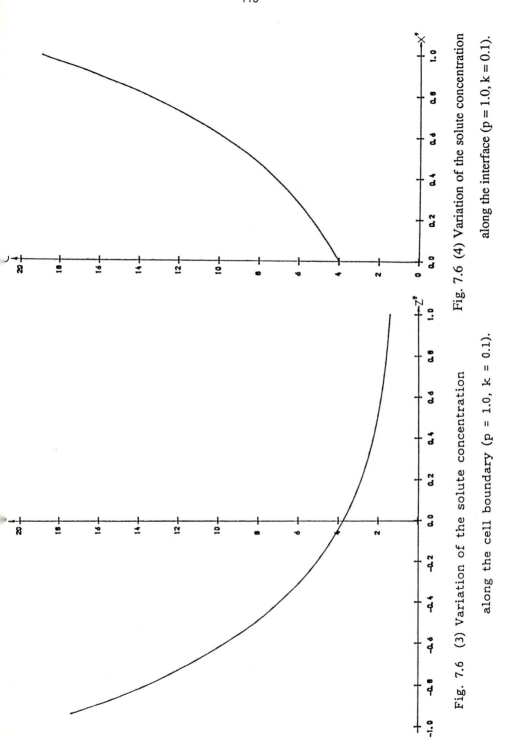

Fig. 7.6 (3) Variation of the solute concentration along the cell boundary (p = 1.0, k = 0.1).

Fig. 7.6 (4) Variation of the solute concentration along the interface (p = 1.0, k = 0.1).

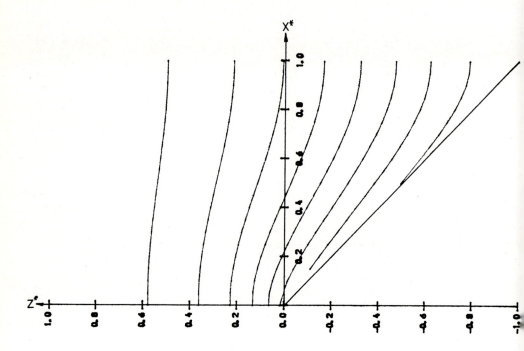

Fig. 7.7 (1) Isoconcentrates (p = 1.0, k = 10.0).

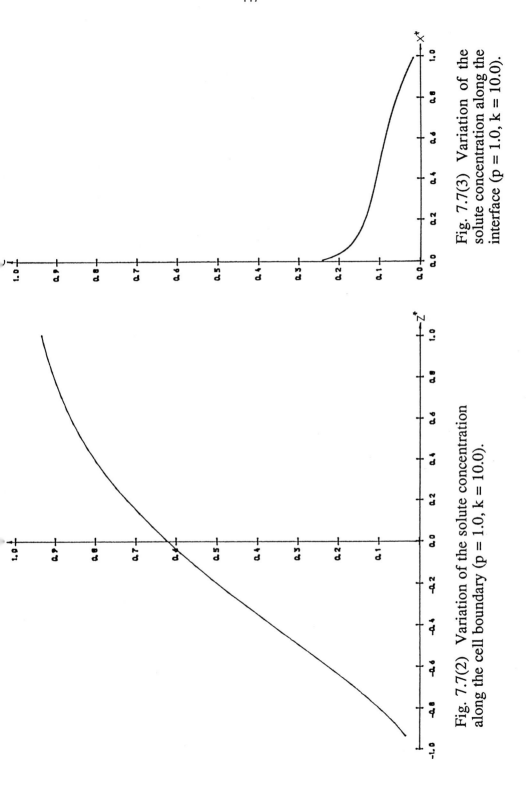

Fig. 7.7(2) Variation of the solute concentration along the cell boundary (p = 1.0, k = 10.0).

Fig. 7.7(3) Variation of the solute concentration along the interface (p = 1.0, k = 10.0).

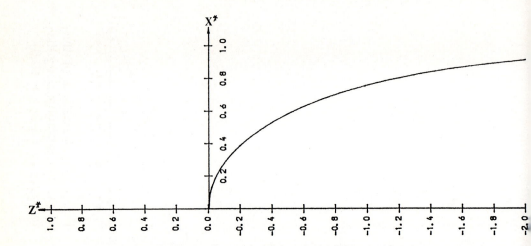

Fig. 7.8 (1) The asymptotic interface.

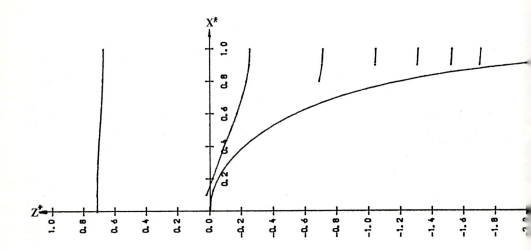

Fig. 7.8 (2) Isoconcentrates (p = 1.0, k = 0.1).

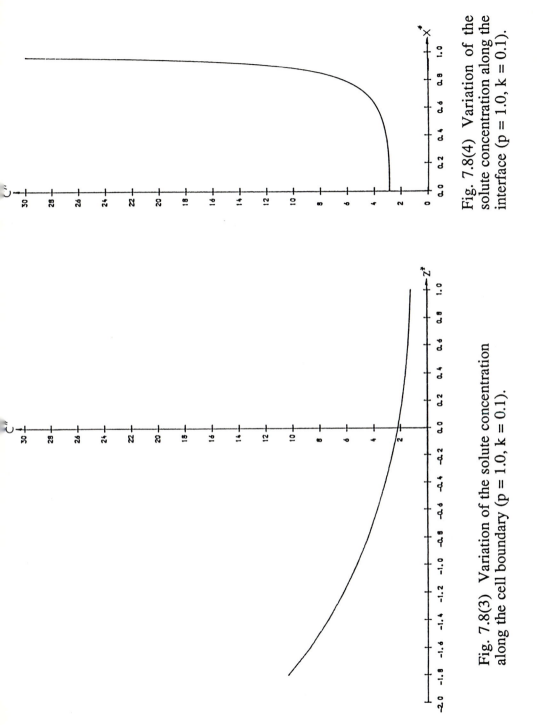

Fig. 7.8(4) Variation of the solute concentration along the interface (p = 1.0, k = 0.1).

Fig. 7.8(3) Variation of the solute concentration along the cell boundary (p = 1.0, k = 0.1).

0.0001 for numerical integration. The computation was performed on a Digital VAX 11/785 computer with double precision which gives a floating point number 8 bytes long having a range of +/-2.9x10^{-37} to +/-1.7x10^{38} and a precision of approximately 16 decimal digits.

Figs. 7.6-7.8 show results (p = 1.0) for the two interface shapes (k = 0.1 for the asymptotic interface, k = 0.1 and k = 10.0 for the faceted interface). Figs. 7.6(4), 7.7(3) and 7.8(4) show the variation of the liquid solute concentration along the interface. It is noted that in the case of k = 0.1, the solute concentration is depressed in the neighbourhood of the tip, *i.e.*, C < CE, where CE is the equilibrium liquid solute concentration at a planar interface; and rises sharply towards the groove, *i.e.*, C > CE. This is to be expected because once solute can only be rejected sideways, the concentration will be determined by a modified Scheil type expression. Figs. 7.6(2), 7.7(1) and 7.8(2) show the isoconcentrates calculated by interpolating between the concentration at grid points linearly along X and exponentially along Z. It can be seen that solute is pronouncedly enriched towards the groove for k = 0.1. Figs. 7.6(3), 7.7(2) and 7.8(3) show the liquid solute concentration along the cell boundary. It can be seen from 7.8(3) that in the groove the concentration varies linearly; this is followed by an exponential decay nearer and ahead of the cell tip. Results for k = 10.0 (Fig. 7.7) demonstrate the solute depletion ahead of the interface for k > 1. However comparison between k = 0.1 and k = 10.0 suggests that these two cases are not symmetric. This is because in the case of k > 1, the solute depletion ahead of the interface will be balanced by the ease of diffusion; while in the case of k < 1 the solute enrichment ahead of the interface will further deteriorate because of the increase of the burden for diffusion, and eventually the solute enrichment will become catastrophic leading to the formation of deep liquid grooves along the cell boundary.

7.7 Discussions

Eqs. (7.71) and (7.72) can be rewritten as

$$C(X^*,Z^*) = C_\infty - \int_0^1 [G(X^*,Z^*,X'^*,Z'^*) \, S(X'^*,Z'^*)]_{Z'^* = \varepsilon(X'^*)} \, dX'^* -$$

$$- \int_0^1 \left[D(X'^*,Z'^*) \frac{\partial G(X^*,Z^*,X'^*,Z'^*)}{\partial n'^*} \right]_{Z'^* = \varepsilon(X'^*)} \left(\frac{ds'^*}{dX'^*} \right) dX'^*$$

(7.95)

with

$$[S(X'^*, Z'^*)]_{Z'^* = \varepsilon(X'^*)} = 2p\left[k\, C(X'^*, Z'^*) - C_\infty\right]_{Z'^* = \varepsilon(X'^*)} \quad (7.96)$$

$$[D(X'^*, Z'^*)]_{Z'^* = \varepsilon(X'^*)} = \left[C(X'^*, Z'^*) - C_\infty\right]_{Z'^* = \varepsilon(X'^*)} \quad (7.97)$$

S(X'*, Z'*) and D(X'*, Z'*) represent a "single layer density" and a "double layer density" respectively. It can be seen from Eq. (7.95) that, the single layer density enters the integral equation as multiplied by the Green's function G, while the double layer density enters the integral equation as multiplied by the normal gradient of the Green's function at the interface, $\frac{\partial G}{\partial n'^*}$. In a special case, *i.e.*, the well-known case of a planar interface, we have

$$[C(X'^*, Z'^*)]_{Z'^* = 0} = \frac{C_\infty}{k} \qquad (7.98)$$

As we can see from Eq. (7.96), the single layer density vanishes, and the concentration solution is represented entirely by the double layer density yielding

$$C(X^*, Z^*) = C(Z^*) = C_\infty + \frac{1-k}{k} C_\infty \exp(-2 p Z^*) \qquad (7.99)$$

which is the well-known solution for a planar interface.

It can be seen that, the solute field is determined by the strength of the source, *i.e.*, the interface concentration, and the Green's function (of strength unity) representing the boundary conditions and the position of the point of observation relative to the point of source.

As has been demonstrated, the technique developed is effective in solving the steady state solute diffusion problem with a moving interface for a given interface shape. Practically under a certain growth condition, the steady state interface shape is determined by heat flow as well as solute flow. Normally the heat flow problem can be simplified by assuming a linear temperature gradient. The interface shape is then determined by the interface undercooling condition,

$$T_0 - T_I = \Delta T_I = m \left(C_\infty - C_I \right) + \theta \left(\frac{1}{R_1} + \frac{1}{R_2} \right) + \frac{V_n}{\mu} \qquad (7.100)$$

To obtain the self-consistent steady state interface shape means finding a solution for $Z = \varepsilon(X)$ such that Eq. (7.100) and all the other boundary conditions are simultaneously satisfied.

7.8 Summary

In this chapter, the point source technique has been presented which solves the steady state solute diffusion problem for a given interface shape, and the solution is expressed in terms of generalized boundary layer potentials. The analytical solution has been satisfactorily evaluated numerically. Numerical experiments have been carried out to evaluate steady state solute fields for different interface shapes under given growth conditions.

Table 7.2 Symbol Table

Symbol	Meaning
C	Solute composition
C_I	Liquid solute concentration at the interface
C_{sI}	Solid solute concentration at the interface
C_∞	Bulk liquid solute composition
D	Solute diffusion coefficient
k	Solute redistribution coefficient
m	Liquidus slope
p	Peclet number
R_1, R_2	Principal radii of curvature
T	Temperature
T_0	Melting temperature
T_I	Interface temperature
ΔT_I	Interface undercooling
V	Steady state growth rate
V_n	Interface normal growth rate
X, Z	Coordinates
θ	Curvature undercooling constant
μ	Kinetic coefficient

* Other symbols used are defined in the text.

Chapter VIII

NUMERICAL DYNAMICAL STUDY OF FACETED CELLULAR ARRAY GROWTH (PART I)

8.1 Introduction

The direct observation of faceted cellular growth, as described in Chapter 5, has clearly revealed the dynamical features of the faceted cellular growth process. As has been demonstrated, there are interactions between cells in an array during the growth process. It is the cellular interaction, through cell tip splitting and loss of cells, which determines the pattern selection of cellular growth. It is therefore essential to consider an array, instead of an isolated cell, in order to describe cellular growth correctly.

The objective of this numerical study, which is to be presented in Chapters 8 & 9, is to understand the physics of the cellular pattern formation process, and to investigate the pattern selection behaviour under different growth conditions. In the work, the time evolution of the faceted cellular interface has been modelled numerically; numerical experiments have been carried out under different growth conditions to examine the pattern selection behaviour. All the terms used in Chapters 8 & 9 together with their units are defined in Table 9.1, unless otherwise indicated.

8.2 The Physical Model

Faceted growth occurs because atomic steps can spread easily over a facet surface, whereas it is difficult to produce new steps (see Chapter 2). So the rate of growth of a facet depends on the rate of production of new steps. It is usually suggested [8] that new steps are produced by two-dimensional nucleation or by an emergent screw dislocation mechanism. With the 2-dimensional nucleation mechanism, the growth rate of a facet is determined by the product of the kinetic undercooling and the facet area. On the other hand, with the screw dislocation nucleation mechanism, the growth rate of a facet is determined by the maximum kinetic undercooling along the facet (see Chapter 2).

Different facet growth rates will lead to interactions between cells in an array. As can be seen from Fig. 8.1, the paths of the joins between facets a, b, and c depend directly on the relevant growth rate of each of the facets in the array; as a consequence, a cell can either become larger or smaller. If a smaller cell becomes larger, the array will be stabilized. On the other hand, if a smaller cell becomes even smaller, it will eventually be grown out by its neighbours in the array, with the neighbouring cells simultaneously becoming larger (Fig. 8.1(1)). However the increase in a cell size will be limited by the fact that the facet can not extend beyond the melting isotherm, because atomic steps can not spread over a part of the facet where there is no kinetic undercooling. Once the local kinetic undercooling at the tip becomes zero, the cell tip will split, and new cells are thus created (Fig. 8.1(2)). As a result of the

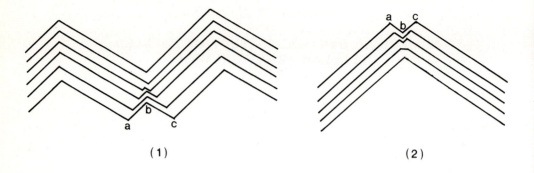

Fig. 8.1 Schametic diagram showing (1) loss of a cell, and (2) cell tip splitting.

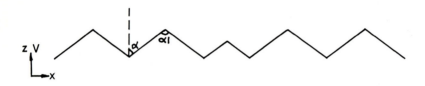

Fig. 8.2 An array of faceted cells.

($a_1 = 2a = 109°28'$)

Fig. 8.3 Growth of the interface over a time step.

cellular interactions, cells are selected in such a way that only those will survive whose spacings lie within a certain range determined by the growth condition.

8.3 The General Scheme of the Numerical Work

The numerical work is aimed at following the time evolution of the faceted cellular interface with the cellular interaction mechanisms incorporated. The numerical models developed are two dimensional, as most of the observations have been made in thin films. However it is thought that the 2-D model preserves the essential features of pattern formation of general faceted cellular growth processes.

Fig. 8.2 shows the typical shape of the faceted cells described in the model. The cells are assumed to be confined by <111> facets, which represent the orientation of facets usually found in Si, Ge, and other f.c.c faceting crystals. The angle between two <111> faces is $109^0 28'$. The choice of the number of cells to start with is made as a compromise between the computational cost and the consideration upon the freedom of the system to allow the cells to interact. Periodical boundary condition is used along both sides of the system, thus allowing the array to extend indefinitely sideways.

The time evolution of the interface is followed explicitly. In a time step, for each facet in the array, the undercooling equation at the interface is first solved

$$\Delta T = \Delta T_k + \Delta T_s + \Delta T_c \qquad (8.1)$$

with

$$\Delta T = G\,Z \qquad (8.2)$$

$$\Delta T_c = 0 \qquad (8.3)$$

Eq. (8.2) implies that a linear temperature gradient is imposed in the liquid ahead of the interface, assuming that the thermal conductivity in both phases is the same and the latent heat evolved is small compared with conducted heat. The effect of growth velocity on the temperature gradient is of second order and thus neglected. Eq. (8.3) means that there is no curvature undercooling for a facet as it is planar.

Once the kinetic undercooling ΔT_k has been calculated from Eq. (8.1), the growth rate of the facets can then be calculated according to the growth kinetics laws. The facets are then moved forward at their growth rates over the time step (Fig. 8.3). The splitting and overgrowth mechanisms as described are incorporated. The time evolution of the interface can then be followed explicitly.

The numerical work starts with an arbitrary law of growth kinetics and with the solute effect neglected (Model I). This sets up the framework for the numerical work. More realistic growth kinetics laws are then introduced (Models II & III). Finally the solute effect is introduced in Model IV. Models I, II & III are described in this chapter while Chapter 9 deals with Model IV and general discussions.

8.4 Model I

8.4.1 Description of the model

This model is in many ways similar to that proposed by Pfeiffer et al. [90]. In this model, it is assumed that, although constitutional undercooling results and cellular growth is initiated, the solute undercooling term does not have significant contribution to the interface undercooling; *i.e.*, the solute undercooling can be neglected. This assumption might be justified if the amount of solute present is just enough for the constitutional undercooling to initiate cellular growth. For the sake of simplicity, in this model it is arbitrarily assumed that, the nucleation rate per unit area, $R_1^{'}$, is proportional to the local kinetic undercooling at the nucleation site, *i.e.*,

$$R_1^{'} = \beta_1 \Delta T_k \tag{8.4}$$

The overall growth rate, R_1, of a facet of area A, is therefore the integration of all the nucleation events along the facet, *i.e.*,

$$R_1 = \int_A \beta_1 \Delta T_{k\,i}\, dA_i \tag{8.5}$$

The kinetic undercooling along each facet in the array can be calculated from the undercooling equations (8.1)--(8.3) with

$$\Delta T_s = 0 \tag{8.6}$$

i.e., the solute undercooling is neglected. Then we have

$$\Delta T_k = \Delta T = G\,(Z_m - Z_i) \tag{8.7}$$

For a facet A (Fig. 8.4), we get

$$R_1 = \int_A \beta_1 \Delta T_{k\,i}\, dA_i = \int_a^b \beta_1\, G\,(Z_m - Z)\frac{dZ}{\cos\alpha} = \beta_1 \frac{G}{2\cos\alpha}(a + b - 2Z_m)(b - a) \tag{8.8}$$

Eq. (8.8) suggests that the growth rate of a facet is proportional to the size of the facet (b - a).

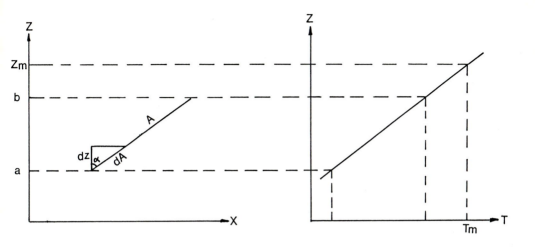

Fig. 8.4 Showing a facet A in the temperature field.

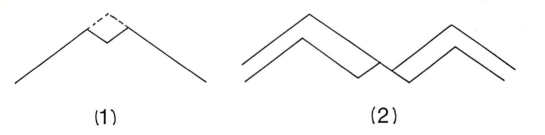

Fig. 8.5 Showing the treatment of (1) tip splitting, and (2) loss of a cell.

8.4.2 Treatment of tip splitting and loss of cells

8.4.2.1 Tip splitting

As was observed experimentally, as a cell tip approaches the melting isotherm, it first flattens along the melting isotherm, then dimpled at the middle, and finally this dimple develops into new faceted cells (see Chapter 5). This sequence is simplified in the model in such a way that a cell splits at its tip instantly once it touches or extends over the melting isotherm (Fig. 8.5(1)), *i.e.*, $\Delta T_k \leq 0$. The newly created facets are given a size of 10^{-11} the original facet size.

8.4.2.2 Loss of cells

During the calculation, the time step is adjusted such that the maximum decrease in a facet size during the time step is just the current facet size; once this happens to a facet, the facet disappears at the end of the time step (Fig. 8.5(2)).

8.4.3 Results

A number of numerical experiments have been carried out to follow the growth process under different conditions. 6 cells with arbitrary perturbations are started with for each calculation. The material parameters are listed in Table 9.2. Fig. 8.6 shows a typical example of the time evolution of the interface. Fig. 8.7 shows the local variation of the average cell spacing with time. Fig. 8.8 shows that cells with different starting spacings approach the same average spacing as the system reaches the steady state. Fig. 8.9 shows the pattern generated by the model compared with the cell boundary network found in Si.

As can be seen from the results, there are strong cellular interactions in the array throughout the growth process. However, as the system settles down into a statistically steady state, the average cell spacing reaches an almost constant value corresponding to the growth condition. Fig. 8.10 shows the steady state average cell spacing calculated for different growth conditions. These results give a relationship

$$\lambda \propto \sqrt{\frac{V}{G}} \tag{8.9}$$

8.4.4 An analytical model of the steady state cell spacing

An approximate form of the model can be obtained by assuming that the average spacing cell has facets large enough to grow at the steady state velocity V (Fig. 8.11), *i.e.*,

$$R = V \sin \alpha \tag{8.10}$$

Assuming that the facet extends up to the melting isotherm, *i.e.*, the cell tip grows with nearly zero kinetic undercooling, gives

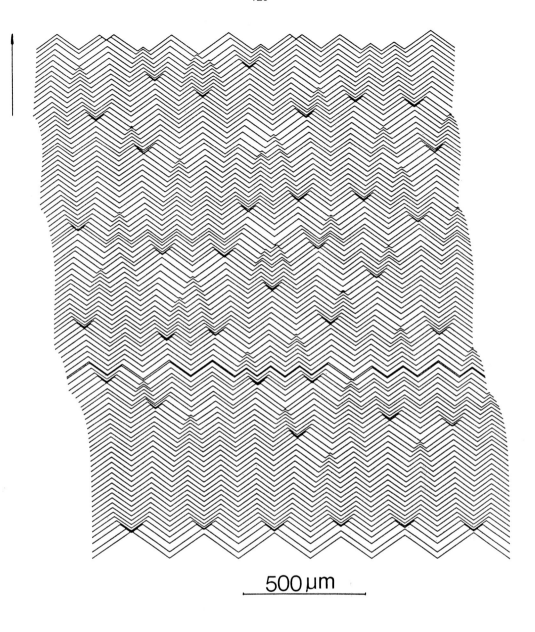

Fig. 8.6 Calculated time evolution of the interface for 2 seconds (V = 1 mm/s, G = 20 K/cm).

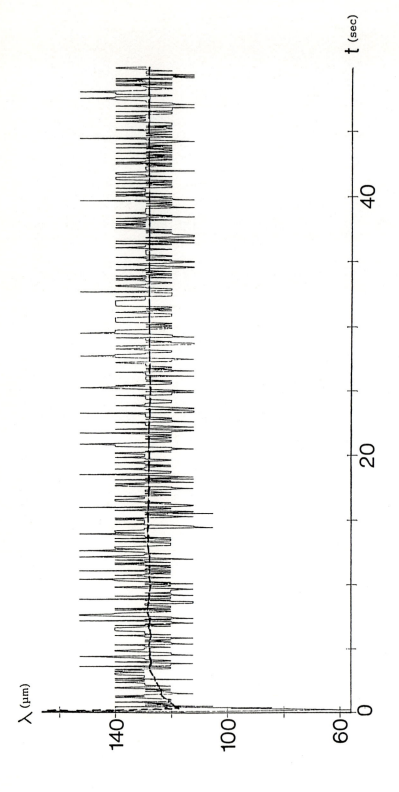

Fig. 8.7 Variation of the cell spacing with time. The solid line represents the local average spacing and the dashed line represents the spacing averaged over time. (V = 1 mm/s, G = 20 K/cm).

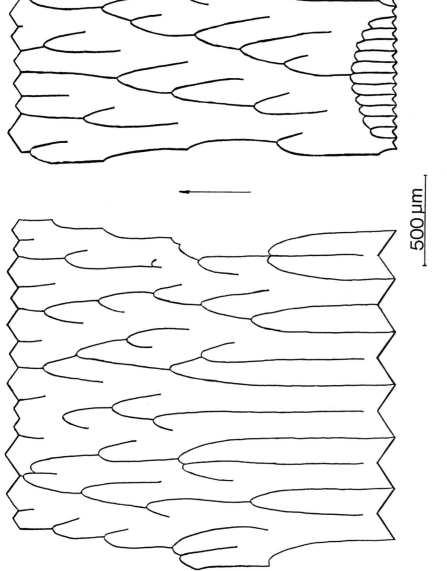

Fig. 8.8 Two arrays with different initial spacings approach the same average spacing after 2 seconds of growth (V = 1 mm/s, G = 20 K/cm).

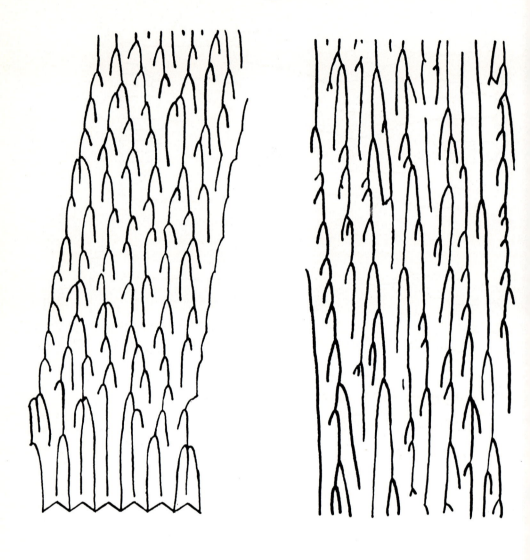

Fig. 8.9 Comparison of computer generated pattern (left) with network in Si (right, taken from Ref. 90).

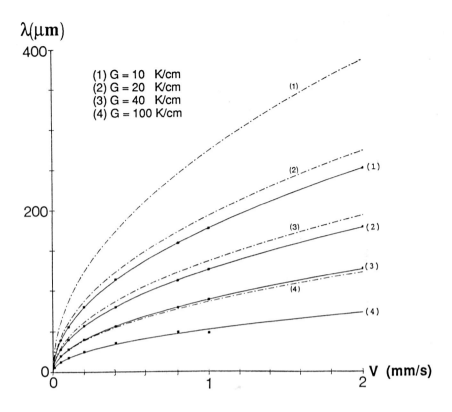

Fig. 8.10 Average cell spacings.
—— Numerical results $\lambda \approx 0.2\,(V/G)^{1/2}$
—·— Analytical model $\lambda \approx 0.27\,(V/G)^{1/2}$

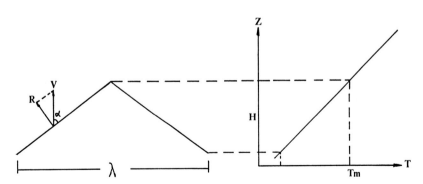

Fig. 8.11 Showing definition of the terms used in Eq. (8.11).

$$R = \int_0^H \beta_1 G (H - Z) \frac{dZ}{\cos \alpha} = \frac{\beta_1 G \lambda^2}{8 \cos \alpha \tan^2 \alpha} \tag{8.11}$$

The spacing then becomes

$$\lambda = 2 \tan \alpha \sqrt{\sin 2\alpha} \sqrt{\frac{V}{\beta_1 G}} \tag{8.12}$$

i.e.,

$$\lambda \propto \sqrt{\frac{V}{G}} \tag{8.13}$$

This is also plotted in Fig. 8.10.

8.4.5 Discussions and Conclusions

The results of the model demonstrates that there are strong, persistent cellular interactions in the array throughout the growth process. This is the result of the facet growth rate being proportional to the facet size. An infinitesimal perturbation can trigger the cellular interaction in the array, and the perturbation will be amplified while being transmitted through the array. However, as a result of this persistent interaction, the system eventually settles down into a statistically steady state when the average cell spacing reaches an almost constant value corresponding to the growth condition.

Both the numerical results and the approximate analytical model give the experimentally measured variation of average cell spacing with velocity [91], $\lambda \propto \sqrt{V}$, and λ decreases with increasing the temperature gradient as expected from the experimental work with thin film Si [93-96] (see Chapter 2).

In conclusion, this simple model clearly demonstrates that the steady state cell spacing is approached through natural selection in an array, without resorting to either the extremum growth criterion or the marginal stability criterion. In other words, the stability concept is incorporated implicitly in the model. It is the cellular interaction which controls pattern selection; therefore it is essential to consider the array in order to describe cellular growth correctly.

8.5 Model II

8.5.1 Description of the model

The previous model is based on Eq. (8.4), which states that the nucleation rate per unit area is proportional to the local kinetic undercooling. This however has little physical significance.

For a two dimensional nucleation mechanism, Eq. (8.4) should be replaced [8] by the following equation for the nucleation rate per unit area, R_2':

$$R_2' = \beta_2 \exp\left(-\frac{K}{\Delta T_k}\right) \qquad (8.14)$$

This is schematically shown in Fig. 8.12. Thus we have

$$R_2 = \int_A \beta_2 \exp\left(-\frac{K}{\Delta T_{k\,i}}\right) dA_i \qquad (8.15)$$

As before an analytical form of the model can be obtained for the cell spacing giving:

$$V = \frac{2\beta_2 K}{(\sin^2 \alpha) G} \Gamma\left[-1, \frac{2 K \tan \alpha}{G \lambda}\right] \qquad (8.16)$$

where Γ is an incomplete Gamma function [118].

8.5.2 Results and discussions

Numerical experiments have been carried out to follow the growth process under a number of growth conditions using a procedure similar to that described for Model I. The material parameters are listed in Table 9.2.

It has been found that, unlike what happens in the previous model, the cells are now much more stable. Under a given growth condition, cells larger than a critical size will split during the transient period and then remain stable. This happens because the cell tip cannot extend beyond the melting isotherm. This gives an upper limit to the cell size. The relationship given by Eq. (8.16) is plotted in Fig. 8.13. The maximum cell spacing is in the same general vicinity.

The stability of this model is the result of the facet growth rate being mainly determined by the kinetic undercooling at the base of the groove and much less by the size of the facet. This is because most of the facet area contributes little to the growth rate, since the exponential term in the nucleation rate equation, Eq. (8.14), has such an effect that a nucleation site with a kinetic undercooling smaller than a critical value has a negligible nucleation rate, as can be seen from Fig. 8.14, in which the variation of the growth rate of a facet is plotted against the size of the facet (assuming that the facet tip has negligible kinetic undercooling). As a consequence, neighbouring cells in the array have nearly identical growth rates and so exhibit little interactions.

8.6 Model III

The previous models should be criticized for the fact that they do not include the possibly more realistic screw dislocation nucleation mechanism, as practically a screw dislocation

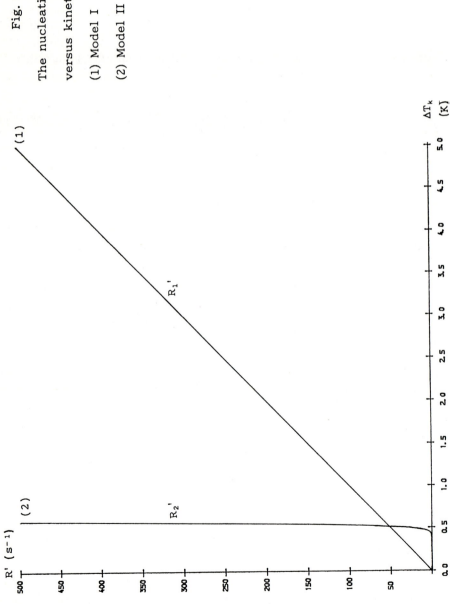

Fig. 8.12

The nucleation rate per unit area versus kinetic undercooling.

(1) Model I
(2) Model II

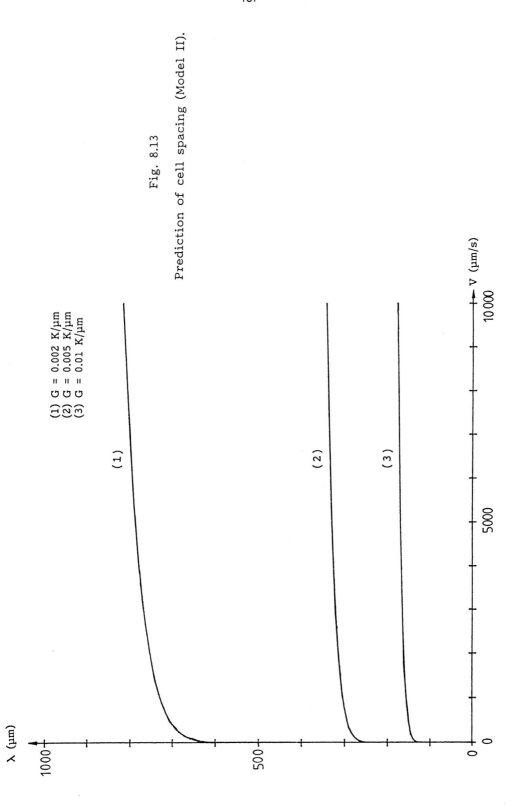

Fig. 8.13 Prediction of cell spacing (Model II).

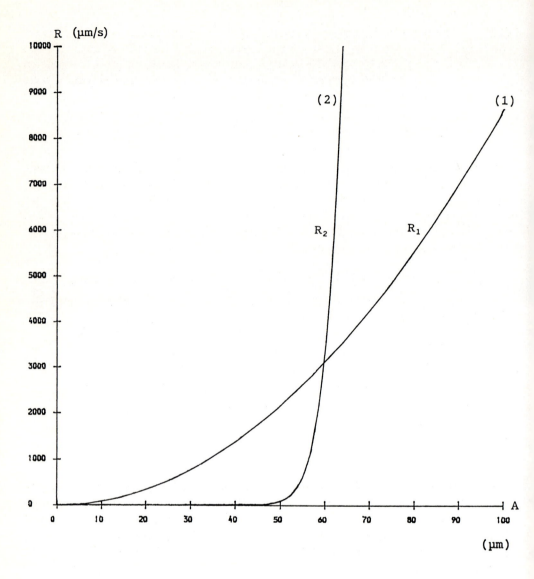

Fig. 8.14 Facet growth rate versus facet size.

(1) Model I

(2) Model II

mechanism should be much more likely than the 2-D nucleation mechanism considered in the previous models for the production of new steps (see Chapter 2). As was mentioned earlier, it has been suggested that, with a screw dislocation nucleation mechanism, the growth rate of a facet is determined by the maximum kinetic undercooling along the facet. It is proposed that the normal growth rate of a facet is given by [8]

$$R_3 = \beta_3 (\Delta T_k)^2_{max} \qquad (8.17)$$

Using similar procedures the time evolution of the interface has been followed. Numerical experiments have shown that stable structures are again produced. This is because the maximum kinetic undercooling along a facet is at the base of the facet, such that two neighbouring facets have the same growth rate, and therefore no interactions should occur. However the fact that the cell tip cannot extend beyond the melting isotherm gives an upper limit to the cell size; cells larger than this will split.

Again an analytical form of the model can be obtained for the maximum cell spacing giving

$$\lambda = 2 \tan \alpha \sqrt{\frac{\sin \alpha}{\beta_3} \frac{\sqrt{V}}{G}} \qquad (8.18)$$

i.e.,

$$\lambda \alpha \frac{\sqrt{V}}{G} \qquad (8.19)$$

8.7 Summary

It has been demonstrated that, Model I, with its arbitrary growth kinetics, produces strong and persistent cellular interactions in the array, resulting in a unique steady state average cell spacing under a given growth condition. Models II & III, with more faithful 2-dimensional nucleation and screw dislocation growth kinetics respectively, produces very stable structures, and an upper limit to the cell size has been found in these two models. Solute effect has been neglected in all of these three models.

Chapter IX

NUMERICAL DYNAMICAL STUDY OF FACETED CELLULAR ARRAY GROWTH
(PART II)

9.1 Introduction

Although the previous models (see Chapter 8) provide a plausible explanation for cellular interactions, it is argued that, they should be criticized for not considering the effect of solute, since under a positive temperature gradient ahead of the interface, cellular growth cannot be initiated without introducing constitutional undercooling; in fact, under normal growth conditions, the solute content is usually considerable [97-98].

In this chapter a numerical model has been developed in which the effect of solute is incorporated and the screw dislocation growth mechanism is included (Model IV). The general scheme of the numerical work has already been described in the previous chapter (see Section 8.3). All the terms used in this chapter together with their units are defined in Table 9.1, unless otherwise indicated.

9.2 Model IV

9.2.1 Description of the model

As was mentioned earlier, with a screw dislocation nucleation mechanism, the growth rate of a facet is determined by the maximum kinetic undercooling along the facet. It is proposed that the normal velocity of a facet is given by [8]

$$R_3 = \beta_3 \left(\Delta T_k\right)^2_{max} \tag{9.1}$$

The effect of solute is incorporated by solving the two dimensional diffusion equation

$$D\left(\frac{\partial^2 C}{\partial X^2} + \frac{\partial^2 C}{\partial Z^2}\right) = \frac{\partial C}{\partial t} \tag{9.2}$$

subject to the far field boundary condition:

$$\left(\frac{\partial C}{\partial Z}\right)_{Z \to \infty} = 0 \tag{9.3}$$

$$C_\infty = C_0 \tag{9.4}$$

and the interface condition:

$$C_{SI} = k\, C_{LI} \qquad (9.5)$$

Eq. (9.5) includes the solute flux equation at the interface. Periodic boundary conditions are used along both sides of the field. Diffusion in the solid is neglected since it is much slower than that in the liquid. The differential equation can be solved by using an explicit control volume finite difference method (see Chapter 3).

Having obtained the composition at a time step, the kinetic undercooling for each facet is calculated from the interface undercooling equations, Eqs. (8.1)-(8.3), with

$$\Delta T_s = m\, C_{LI} \qquad (9.6)$$

The growth rate of each facet is then calculated from Eq. (9.1). The time evolution of the growth process can be followed explicitly in much the same way as in the previous models.

Two cases have been considered: $k > 1$ and $k < 1$. These two cases must be considered separately because they are not symmetric as far as solute redistribution is concerned. As has been demonstrated in Chapter 7, in the case of $k < 1$, solute enrichment at the cell boundary will further increase the burden for diffusion such that the solute enrichment will be further enhanced. The solute concentration can rise up to the eutectic composition. Eventually this situation may become catastrophic leading to the formation of deep liquid grooves along the cell boundary, and the liquid groove will go back up to the eutectic temperature. This however is not the case for $k > 1$, where the solute depletion at the cell boundary will be balanced by the ease of diffusion due to solute depletion; moreover, the depletion of solute will have to stop before the solute concentration reaches 0. Therefore liquid grooves will not normally form in the case of $k > 1$ and shallow cells will be retained.

9.2.2 The modelling procedure

As shown in Fig. 9.1, the calculation starts with a given interface shape with initial solute/temperature fields. The cells grow over a time step at their growth rates calculated. The solute diffusion is solved using a control volume finite difference method, and the new solute/temperature field can be obtained. This is again the starting point for a new time step.

9.2.3 The initial condition

10 identical symmetrical cells are started with for each calculation (Fig. 9.2). No external perturbations are intentionally introduced except numerical discretization errors and computational rounding errors; *i.e.*, the noise level is very low. The initial solute field is as follows: the interface composition is the same as that for an otherwise planar interface under steady state, *i.e.*,

$$C_I = \frac{C_0}{k} \qquad (9.7)$$

Away from the interface, C decays exponentially into the liquid approaching C_0 at infinity, *i.e.*,

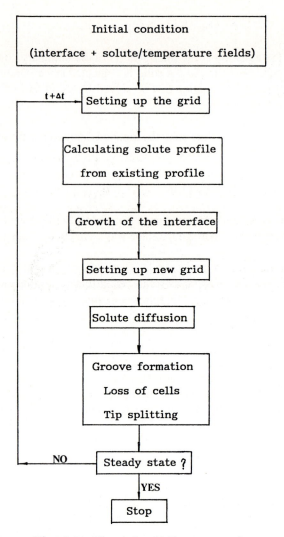

Fig. 9.1 The calculation procedure.

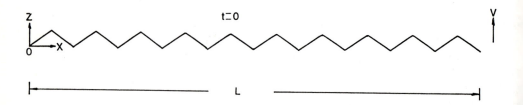

Fig. 9.2 The starting interface shape.

$$C(X, Z) = C(Z) = C_0\left[1 + \frac{1-k}{k}\exp\left(-\frac{VZ}{D}\right)\right] \qquad (9.8)$$

The interface is placed in the temperature field such that the minimum kinetic undercooling, which is at the cell tip, is just above 0. This is so in order to ensure that no dramatic interactions occur during the transient period.

9.2.4 The grid

In order to solve the differential equation, Eq. (9.2), using an explicit control volume finite difference method, it is necessary to divide the whole domain into control volumes. In view of the zigzag feature of the interface, a non-uniform grid is employed (Fig. 9.3). Stationary, Cartesian coordinates are used with the Z axis parallel to the growth direction. The grid is such that grid points lie halfway between box walls, and the interface lies across the grid points. The width of the box near the interface, d, is of the order of (0.1 D/V) depending on the size of the facet such that the number of boxes for each facet is 10 to 20. The considerations for the box size include: (1) The box size should not be too large in order to ensure numerical accuracy; (2) The number of boxes for a facet of average size should not be too few (at least 10 were retained in the calculations); (3) Considerations upon the computational cost: more boxes mean more calculations to perform; also smaller box sizes require smaller time step in order to ensure mathematical stability (see 9.2.11.1).

As the facet sizes are different across the interface, the box width d_M is only uniform around each facet region, with $d \leq d_M \leq 2d$. Away from the interface the solute field tend to be uniform; thus the box height is increased by a factor of 2 each time. The modelled domain extends into the liquid up to the distance of the whole interface width L, or (D/V), whichever is the largest. In most cases around 300 x 30 boxes are used.

9.2.5 Growth of the cells

Along each facet, the kinetic undercooling at each grid point is calculated from Eq. (8.1), and the maximum kinetic undercooling along the facet can be obtained. The growth rate of the facet is then calculated from Eq. (9.1). The facets across the interface are moved forward at their growth rates over the time step Δt (Fig. 9.4). Periodic boundary conditions must be taken into account. The temperature field is at the same time moving at a constant velocity V. The northern border of the modelled liquid region also moves by $V \Delta t$ (Fig. 9.4).

9.2.6 The grid for the new interface

A facet normally changes its position and size after growth. At each new time step, the facet is again uniformly divided into the same number of boxes as the original facet. As before the box height in the liquid region increases by a constant factor such that the same number of boxes covers the whole liquid region to be modelled as in the original row.

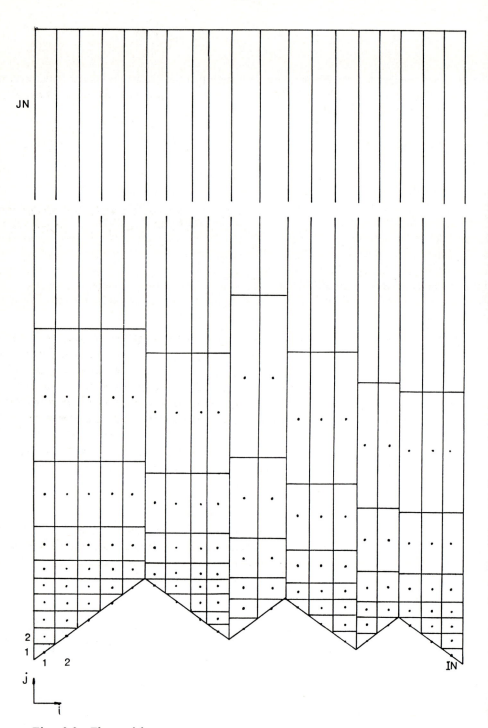

Fig. 9.3 The grid.

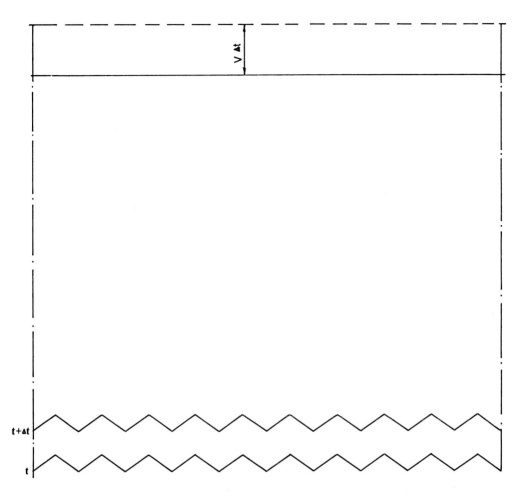

Fig. 9.4 Growth of the interface.

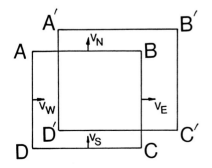

Fig. 9.5 A control volume ABCD.

S = ABCD, S' = A'B'C'D'.

9.2.7 Solution of solute diffusion during growth

The diffusion equation, Eq. (9.2), is solved with an explicit control volume finite difference method. For a control volume ABCD (Fig. 9.5), the discretisation equation is

$$\frac{\partial (C\,S)}{\partial t} = Q \tag{9.9}$$

Thus we have

$$(C'\,S') - (C\,S) = Q\,\Delta t \tag{9.10}$$

$$C' = \frac{Q\,\Delta t + C\,S}{S'} \tag{9.11}$$

Q is the total flux flowing into the box (Fig. 9.6), *i.e.*,

$$Q = \sum_i Q_i \qquad i = E, W, N, S \tag{9.12}$$

where E, W, N, and S denote the east, west, north, and south, respectively, and

$$Q_E = \left(D\,\frac{C_{i+1,j} - C_{i,j}}{\Delta X_E} + V_E\,\Delta t\,C_E \right) \Delta Z \tag{9.13}$$

$$Q_W = \left(D\,\frac{C_{i-1,j} - C_{i,j}}{\Delta X_W} - V_W\,\Delta t\,C_W \right) \Delta Z \tag{9.14}$$

$$Q_N = \left(D\,\frac{C_{i,j+1} - C_{i,j}}{\Delta Z_N} + V_N\,\Delta t\,C_N \right) \Delta X \tag{9.15}$$

$$Q_S = \left(D\,\frac{C_{i,j-1} - C_{i,j}}{\Delta Z_S} - V_S\,\Delta t\,C_S \right) \Delta X \tag{9.16}$$

with

$$V_E\,\Delta t = X_{B'C'} - X_{BC} \tag{9.17}$$

$$V_W\,\Delta t = X_{A'D'} - X_{AD} \tag{9.18}$$

$$V_N\,\Delta t = Z_{A'B'} - Z_{AB} \tag{9.19}$$

$$V_S\,\Delta t = Z_{D'C'} - Z_{DC} \tag{9.20}$$

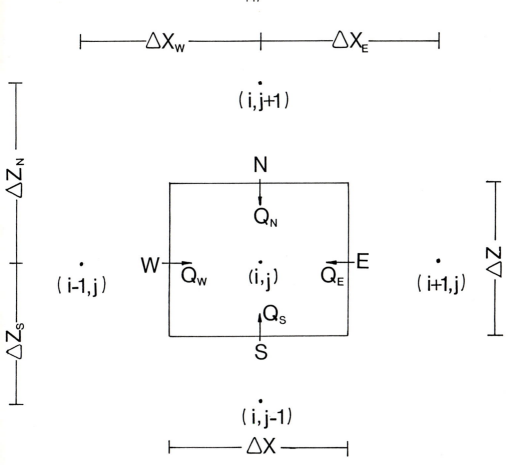

Fig. 9.6 Flux balance over a control volume.

C_E and C_W can be calculated by linearly interpolating along X between the adjacent points of the eastern and western walls respectively (Fig. 9.7), which gives

$$C_E = C_{i,j} + \frac{C_{i+1,j} - C_{i,j}}{\Delta X_E} \frac{\Delta X}{2} \quad (9.21)$$

$$C_W = C_{i,j} + \frac{C_{i-1,j} - C_{i,j}}{\Delta X_W} \frac{\Delta X}{2} \quad (9.22)$$

C_N and C_S can be calculated by exponentially interpolating along Z between the adjacent points of the northern and southern walls respectively (Fig. 9.8), as this is the solution of the corresponding one dimensional steady state diffusion problem with a moving interface. This gives

$$C_N = C_{i,j} + B_N \left[\exp\left(-\frac{V}{D} \frac{\Delta Z}{2}\right) - 1 \right] \quad (9.23)$$

$$C_S = C_{i,j} + B_S \left[\exp\left(\frac{V}{D} \frac{\Delta Z}{2}\right) - 1 \right] \quad (9.24)$$

with

$$B_N = \frac{C_{i,j} - C_{i,j+1}}{1 - \exp\left(-\frac{V \Delta Z_N}{D}\right)} \quad (9.25)$$

$$B_S = \frac{C_{i,j} - C_{i,j-1}}{1 - \exp\left(\frac{V \Delta Z_S}{D}\right)} \quad (9.26)$$

Applying appropriate boundary conditions (except the formation and treatment of liquid grooves which will be considered separately in (9.2.8)) leads to the following equations:

(1) The far field condition (j = JN)

From Eqs. (9.3), (9.4) and (9.15) we have

$$Q_{N(j=JN)} = V \Delta t \, C_0 \, \Delta X \quad (9.27)$$

(2) The interface condition (j = 1) (Fig. 9.9)

From Eq. (9.5) we can get

$$Q_I = R \, \Delta t \, \Delta L \, k \, C_{i,j} \quad (9.28)$$

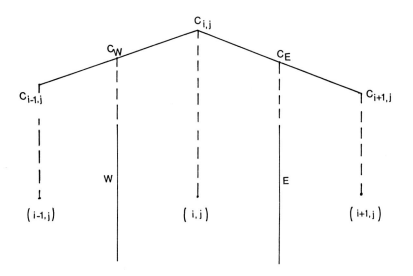

Fig. 9.7 Linear interpolation.

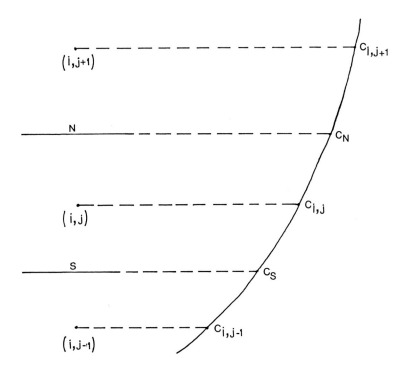

Fig. 9.8 Exponential interpolation.

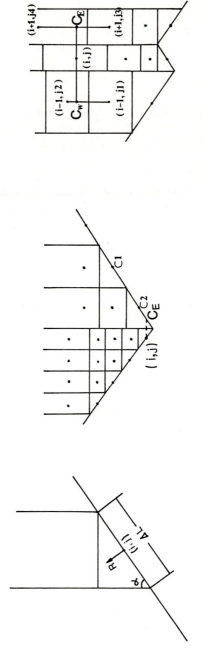

Fig. 9.9 An interface box.

Fig. 9.10 An interface box at the base of the cell.

Fig. 9.11 Boxes adjacent to another facet.

with
$$\Delta L = \sqrt{(\Delta x)^2 + (\Delta z)^2} \tag{9.29}$$

where

$$Q_I = \begin{cases} Q_S + Q_E & \text{for a LHS facet} \\ Q_S + Q_W & \text{for a RHS facet} \end{cases} \tag{9.30}$$

(3) Boxes at the base of the cell (Fig. 9.10)

C_E (or C_W for a LHS facet) is calculated by linearly interpolating between interface points along the adjacent facet, C_1 and C_2.

(4) Boxes adjacent to another facet (Fig. 9.11)

These boxes cannot find an immediate eastern (or western) neighbour point since they are not lined up with boxes around the adjacent facet. Therefore for these boxes, C_W (or C_E) is calculated by exponentially interpolating along Z between nodes (i-1, j1) and (i-1, j2) (or (i+1, j3) and (i+1, j4) for C_E).

(5) The periodic boundary condition (i = 1, i = IN) (Fig. 9.12)

The periodic boundary condition means that the field modelled continues along X from its own end, *i.e.*,

For i = 1,

$$C_{i-1,j} = C_{IN,j}$$

$$Z_{i-1,j} = Z_{IN,j}$$

$$X_{i-1,j} = X_{IN,j} - L \tag{9.31}$$

For i = IN,

$$C_{i+1,j} = C_{1,j}$$

$$Z_{i+1,j} = Z_{1,j}$$

$$X_{i+1,j} = X_{1,j} + L \tag{9.32}$$

Finally Eq. (9.11) can be solved one by one for all the boxes to get the solute field.

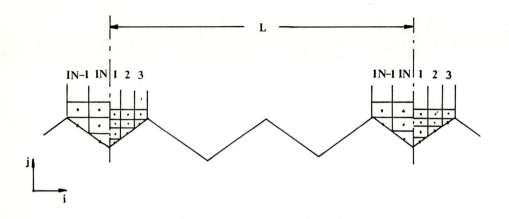

Fig. 9.12 The periodic boundary condition.

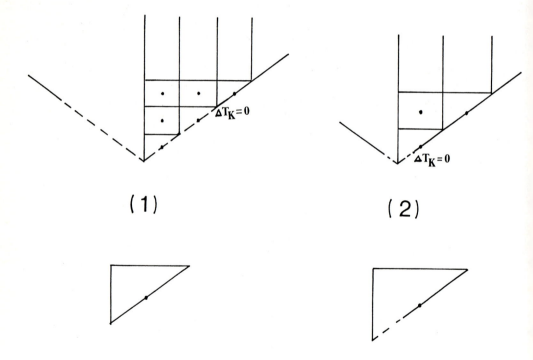

Fig. 9.13 Boxes at the liquid groove.

(1) A purely liquid box.

(2) A composite box.

9.2.8 Formation and treatment of liquid grooves

As was mentioned earlier, liquid grooves may form along cell boundaries as a result of severe solute build-up. Once formed, it will serve as a channel for solute flow. The liquid groove is normally non-faceted and should follow the isotherm $\Delta T_k = 0$ (assuming $\Delta T_k = 0$ for non-faceted growth). However the shape of the liquid groove is not modelled in this work; instead only its contribution to solute flow is incorporated.

9.2.8.1 Position of the liquid groove

By linearly interpolating the composition (or kinetic undercooling, as the temperature field is linear) between grid points along a facet, the position of the point near the base of the facet can be calculated at which $\Delta T_k = 0$ (if it exists). This position is taken as the starting point of the liquid groove (Fig. 9.13). The width of a groove is assumed to be unchanged during a time step; at the end of the time step it is recalculated as was described.

9.2.8.2 Solute flow along the groove

Originally only one box was used for the groove. It was found that large grooves were formed which were almost of the same width as the half cell spacing, and the size of the groove could not change once formed. Obviously this does not seem to be a realistic situation. An improvement has been made such that the number of boxes within a groove varies with the size of the groove. If the groove width is larger than d, one or more boxes (purely liquid) are used (Fig. 9.13(1)). However if the groove width is smaller than d, a composite box, partially liquid and partially solid, is used (Fig. 9.13(2)).

Along the liquid groove, it is thought that there is sufficient time for the liquid to become homogeneous in the X direction. Neglecting the curvature and kinetic undercooling along the groove, we get [41,42]

$$G\,dZ = dT = m\,dC \tag{9.33}$$

Thus

$$\frac{dC}{dZ} = \frac{G}{m} \tag{9.34}$$

The flux term across the wall can then be calculated. For a composite box, however, the flux is the sum of the flow across the solid part of the wall, to be calculated from Eq. (9.28), and the flow across the liquid part of the wall, to be calculated from Eq. (9.34). This treatment has produced grooves which seem to agree with the experimental observations (see 9.3).

9.2.9 Treatment of tip splitting and loss of cells

9.2.9.1 Loss of cells

It is assumed that a facet, once becoming smaller than a critical size, which is chosen to be d/2 in the present work, disappears instantly.

9.2.9.2 Tip splitting

This is done in the same way as was described in the previous models, *i.e.*, once ΔT_k becomes 0 at a cell tip, the cell tip is assumed to split instantly. The initial size of a newly created facet is about the box size d. It is also assumed that the new facets cannot grow immediately after being created. A composition is given to the new facet such that it will have a negligible kinetic undercooling. The composition of its immediate neighbour in the liquid is recalculated by exponentially interpolating along Z, in order to smooth out any distortion this may have caused to the solute field in the neighbourhood.

9.2.10 Calculation of the solute profile at new grid points

After growth, both the number of cells and the sizes of the facets have changed, and therefore the grid is rearranged as described in 9.2.4. Compositions for the new grid points are calculated from the existing solute profile by interpolation, *i.e.*, the new solute profile is mapped from the existing solute profile.

See Fig. 9.14. Compositions at points CA, CB are first calculated by exponentially interpolating between C1 and C2, C3 and C4 respectively; then by linearly interpolating between CA and CB, the composition of the new grid point CN is obtained. The composition of interface points are calculated by linearly interpolating between the existing interface points. Periodic boundary conditions must be taken into account in the calculation.

9.2.11 Computational considerations

9.2.11.1 Choice of the time step

Considerations include:
(1) The stability criterion for an explicit finite difference scheme (see Chapter 3) requires:

$$\Delta t \le (\Delta t)_c = \frac{(d_m)^2}{4D} \tag{9.35}$$

where d is the minimum box size. As can be seen, the present model involves the loss and creation of boxes, loss and creation of facets, during the calculation, and much work with interpolations. Moreover, the model does not start with a steady state solution. All this makes the calculation more vulnerable to mathematical instabilities. In order to ensure mathematical stability, the calculation starts with a time step of $(0.1\ \Delta t_c)$; then the time step is gradually increased to Δt_c after about 10000 cycles.

(2) The time step should be small enough to resolve the time evolution of the interface;

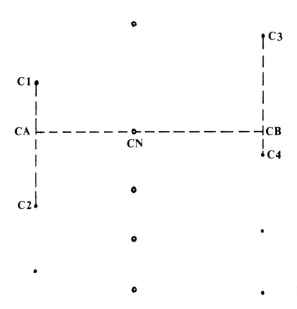

Fig. 9.14 Interpolation of the solute field.

(3) The growth of a facet is confined within a box during a time step; and

(4) The decrease in a facet size during a time step should not exceed the current facet size.

9.2.11.2 Monitoring the calculation

After solving the diffusion problem (see 9.2.7), and after calculating the new solute profile from the existing solute profile by interpolations (see 9.2.10), mass conservation is checked. The variation of average solid concentration with time is also checked to see if there are any significant mathematical oscillations.

9.2.11.3 The computational efforts

The computation has been performed on a Compaq 386/20™ microcomputer with double precision, which may represent values in the range 9.46×10^{-308} - 1.80×10^{308}, precise to 15 digits.

9.3 Results and Discussions

In spite of the comparatively large computational time required, a number of numerical experiments have been carried out under different growth conditions. The material parameters used in the calculation are listed in Table 9.2.

9.3.1 Faceted cellular array growth

Results of the numerical experiments are listed in Table 9.3 (for $k < 1$) and Table 9.4 (for $k > 1$). In each experiment, the time evolution of the interface has been calculated for different initial cell spacings under the same growth condition. Figs. 9.15 - 9.22 show examples of the calculated time evolution of the interface for numerical experiments No. 1 - No. 8 as listed in Tables 9.3 & 9.4. For clarity of presentation, only some of the calculated interface positions were shown, and the time interval between two consecutive interface positions shown in the figures do not correspond to the actual time step used in the numerical calculation.

As can be seen, under a given growth condition, there is a range of stable cell spacings, below which loss of cells lead to an increase in the cell spacing, and above which the creation of new cells through tip splitting leads to a decrease in the cell spacing. The cells remain stable after the transient interactions in the array. Eventually the cellular interface approaches a steady state.

The initial cell spacing and the final cell spacing for each experiment are summarized in Fig. 9.23, which shows that under a given growth condition, the final cell spacings all fall into the stable range, whatever the initial cell spacings are.

Table 9.1 Symbol Table (Chapters 8&9)

Symbol	Meaning	Unit
A	Facet area	µm
C	Solute concentration	Atomic fraction
C_0	Bulk liquid solute concentration	Atomic fraction
C_{LI}	Liquid solute concentration at the interface	Atomic fraction
C_{SI}	Solid solute concentration at the interface	Atomic fraction
D	Solute diffusivity in the liquid	$\mu m^2 \, s^{-1}$
G	Temperature gradient in the liquid	$K \, \mu m^{-1}$
k	Solute redistribution coefficient	
K	Kinetic coefficient (Model II)	K
m	Liquidus slope	$K \, (\text{atomic fraction})^{-1}$
Q	Solute flux	
R	Facet normal growth rate	$\mu m \, s^{-1}$
R_1	Facet normal growth rate (Model I)	$\mu m \, s^{-1}$
R_2	Facet normal growth rate (Model II)	$\mu m \, s^{-1}$
R_3	Facet normal growth rate (Models III&IV)	$\mu m \, s^{-1}$
R'_1	Nucleation rate per unit area (Model I)	s^{-1}
R'_2	Nucleation rate per unit area (Model II)	s^{-1}
t	Time	s
T	Temperature	K
ΔT	Interface undercooling	K
ΔT_c	Curvature undercooling	K
ΔT_k	Kinetic undercooling	K
ΔT_s	Solute undercooling	K
V	Velocity	$\mu m \, s^{-1}$
X, Z	Coordinates	µm
β_1	Kinetic coefficient (Model I)	$s^{-1} \, K^{-1}$
β_2	Kinetic coefficient (Model II)	s^{-1}
β_3	Kinetic coefficient (Models III&IV)	$\mu m \, s^{-1} \, K^{-2}$
λ	Cell spacing	µm
Γ	Gamma function	

* Other symbols used in Chapters 8&9 are defined in the text.

Table 9.2   Material Parameters

β_1	β_2	β_3	D	K	m (k>1)	m (k<1)
100	10^{10}	10,000	3,500	10	500	-500

Table 9.3 Results of Numerical Experiments ($k = 0.2$, $C_0 = 0.02\%$)

Experiment No.	Run No.	G (K/µm)	V (µm/s)	λ (µm) Initial	λ (µm) Final	Stable cell spacing (µm) λ_{min}	Stable cell spacing (µm) λ_{max}
1	1	0.005	45	1	1	1	150
	2			5	5		
	3			30	30		
	4			40	40		
	5			50	25		
	6			70	35		
	7			150	75		
	8			300	150		
	9			400	150		
	10			500	128		
2	11	0.005	100	10	50	40	167
	12			20	40		
	13			50	100		
	14			60	120		
	15			70	70		
	16			80	80		
	17			100	100		
	18			160	160		
	19			200	167		
	20			250	125		
	21			260	130		
	22			270	135		
	23			300	150		

Table 9.3 (continued)

Experiment No.	Run No.	G (K/μm)	V (μm/s)	λ (μm) Initial	λ (μm) Final	Stable cell spacing (μm) λ_{min}	Stable cell spacing (μm) λ_{max}
3	24	0.005	500	20	67	50	182
	25			30	60		
	26			40	50		
	27			60	60		
	28			100	100		
	29			140	140		
	30			150	150		
	31			160	160		
	32			170	121		
	33			180	90		
	34			200	100		
	35			300	150		
	36			350	175		
	37			400	182		
	38			600	120		
4	39	0.005	1,000	20	40	40	190
	40			30	60		
	41			40	40		
	42			160	160		
	43			180	180		
	44			190	190		
	45			200	167		
	46			230	115		
	47			300	150		
	48			400	133		
	49			500	125		
	50			600	130		

Table 9.3 (continued)

Experiment No.	Run No.	G (K/μm)	V (μm/s)	λ (μm) Initial	λ (μm) Final	Stable cell spacing (μm) λ_{min}	Stable cell spacing (μm) λ_{max}
5	51	0.005	10,000	5	17	11	500
	52			10	11		
	53			20	20		
	54			50	50		
	55			120	120		
	56			190	190		
	57			500	500		
	58			600	150		
	59			800	133		
6	60	0.05	500	1	1	1	32
	61			10	10		
	62			30	15		
	63			50	25		
	64			100	32		
	65			150	21		
7	66	0.0005	500	40	100	50	800
	67			50	50		
	68			100	100		
	69			500	500		
	70			800	800		
	71			900	600		
	72			1,000	450		

Table 9.4 Results of Numerical Experiments ($k = 5.0$, $C_0 = 0.05\%$)

Experiment No.	Run No.	G (K/µm)	V (µm/s)	λ (µm) Initial	λ (µm) Final	Stable cell spacing (µm) λ_{min}	Stable cell spacing (µm) λ_{max}
8	73	0.001	100	50	100	70	300
	74			70	70		
	75			100	100		
	76			200	200		
	77			300	300		
	78			350	175		
	79			400	200		
	80			500	250		

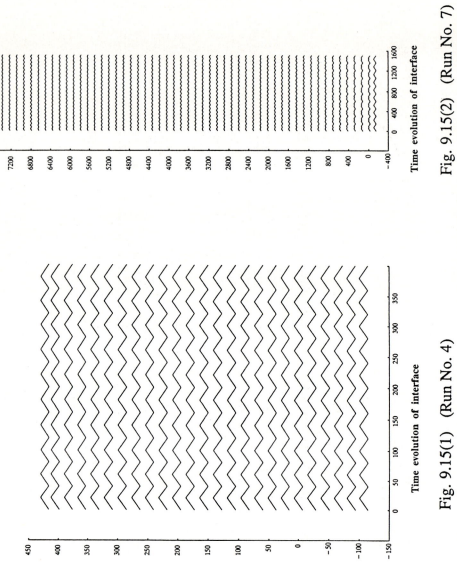

Fig. 9.15 Time evolution of the interface (Numerical experiment No. 1). (μm)

Fig. 9.15(1) (Run No. 4)

Fig. 9.15(2) (Run No. 7)

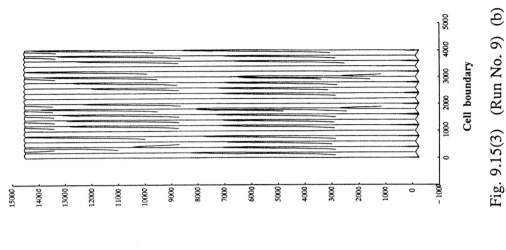

Fig. 9.15(3) (Run No. 9) (a) Time evolution of interface

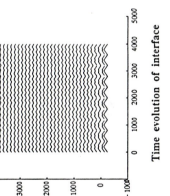

Fig. 9.15(3) (Run No. 9) (b) Cell boundary

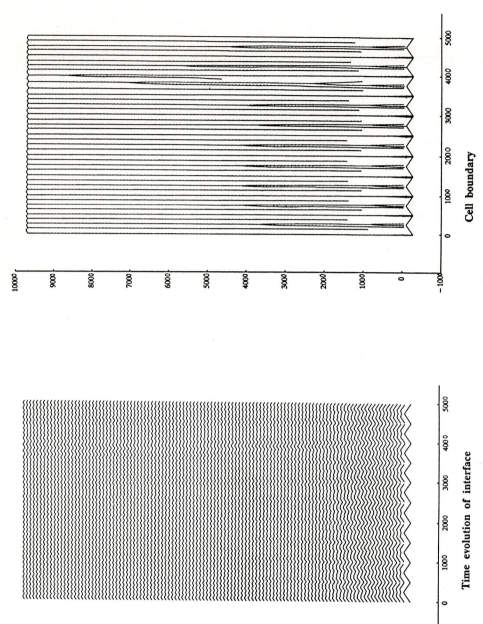

Fig. 9.15(4) (Run No. 10) (a) Time evolution of interface (b) Cell boundary

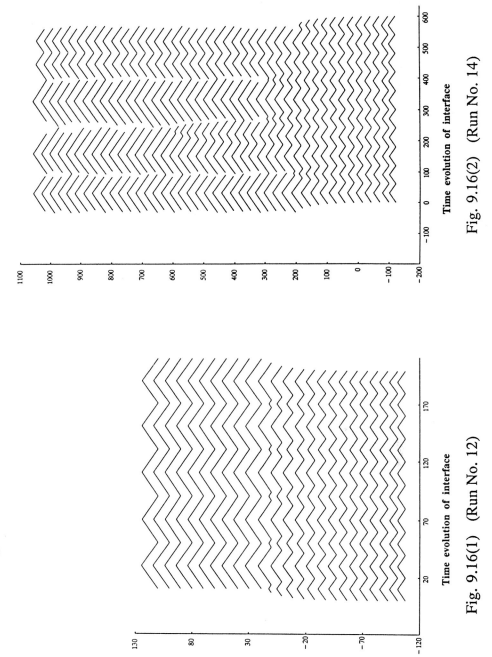

Fig. 9.16 Time evolution of the interface (Numerical experiment No. 2). (μm)

Fig. 9.16(1) (Run No. 12)

Fig. 9.16(2) (Run No. 14)

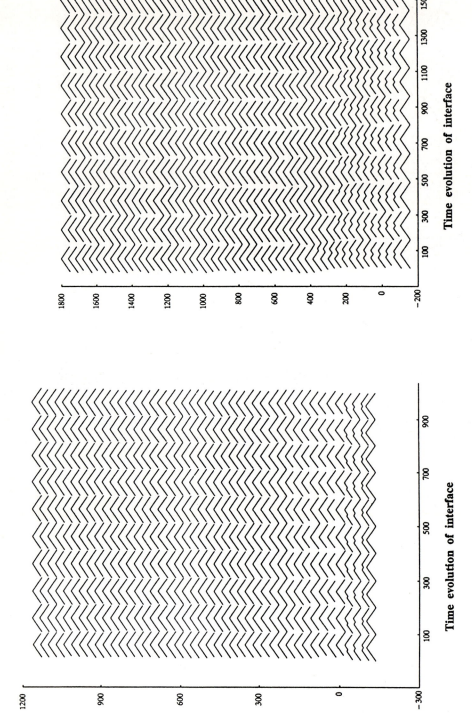

Fig. 9.16(3) (Run No. 17)

Fig. 9.16(4) (Run No. 18)

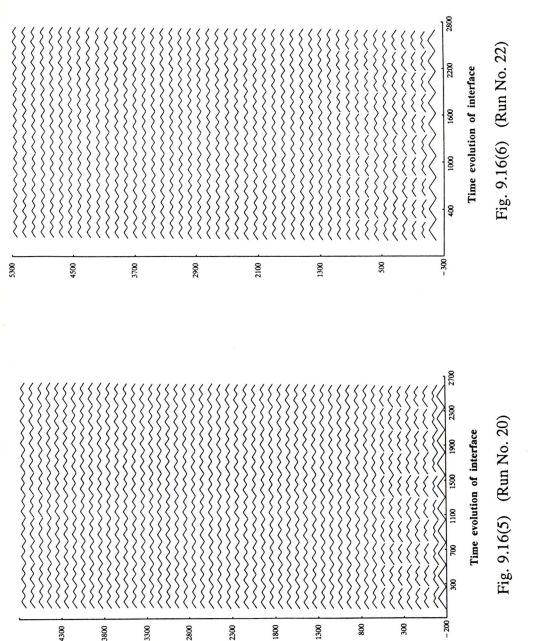

Fig. 9.16(5) (Run No. 20)

Fig. 9.16(6) (Run No. 22)

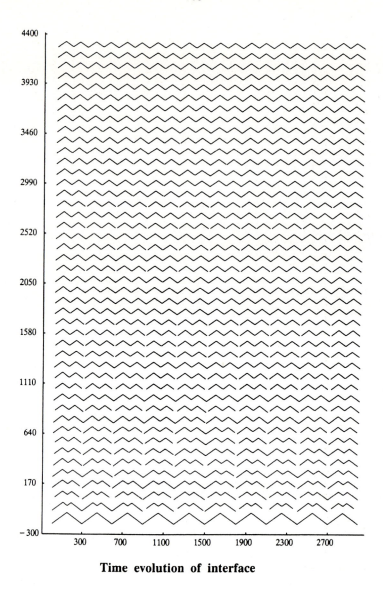

Time evolution of interface

Fig. 9.16(7)   (Run No. 23)

Fig. 9.17 Time evolution of the interface (Numerical experiment No. 3). (μm)

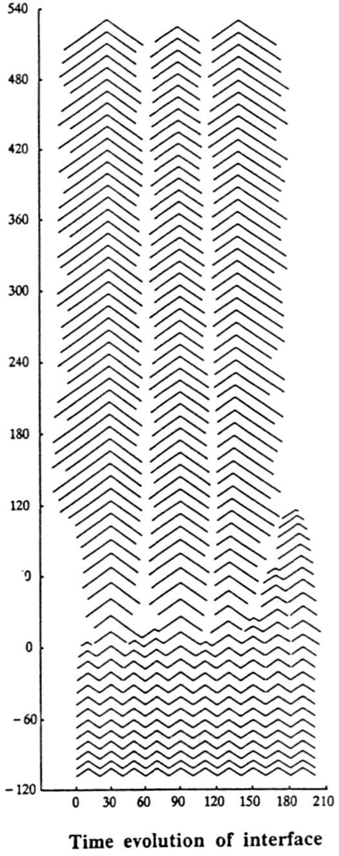

Fig. 9.17(1) (Run No. 24)

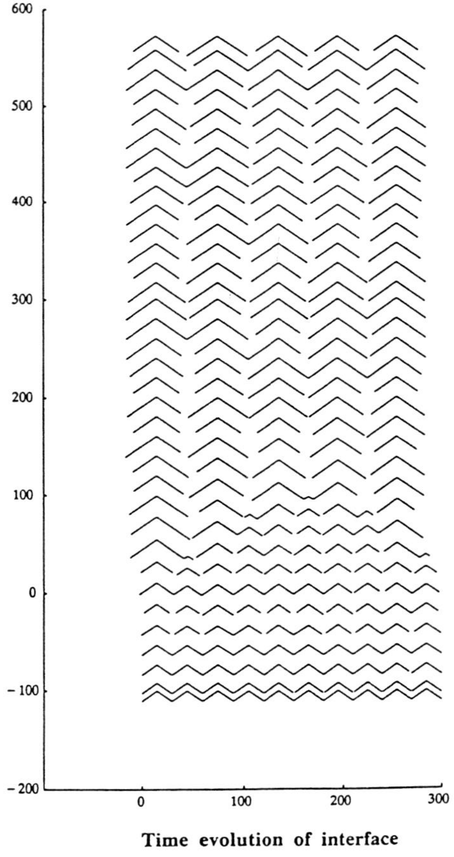

Fig. 9.17(2) (Run No. 25)

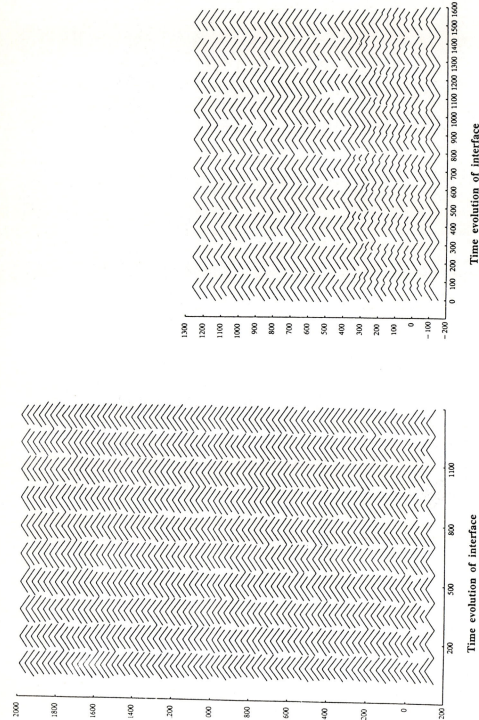

Fig. 9.17(3) (Run No. 29)

Fig. 9.17(4) (Run No. 31)

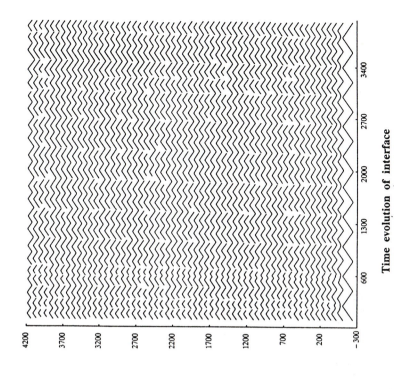

Time evolution of interface

Fig. 9.17(6) (Run No. 37)

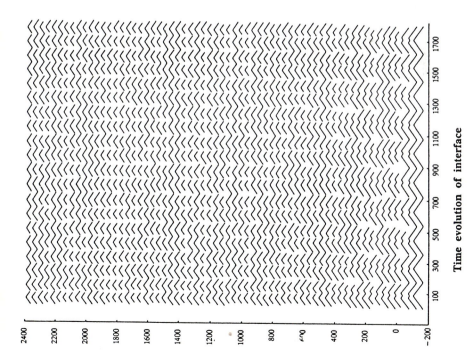

Time evolution of interface

Fig. 9.17(5) (Run No. 33)

Fig. 9.18 Time evolution of the interface (Numerical experiment No. 4). (μm)

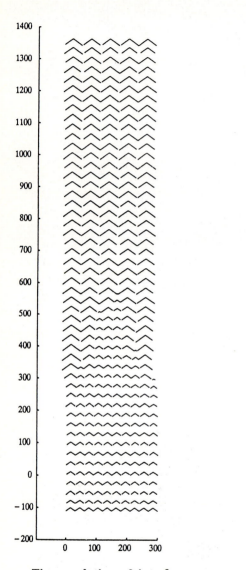

Time evolution of interface

Fig. 9.18(1) (Run No. 40)

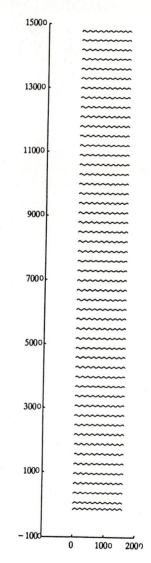

Time evolution of interface

Fig. 9.18(2) (Run No. 42)

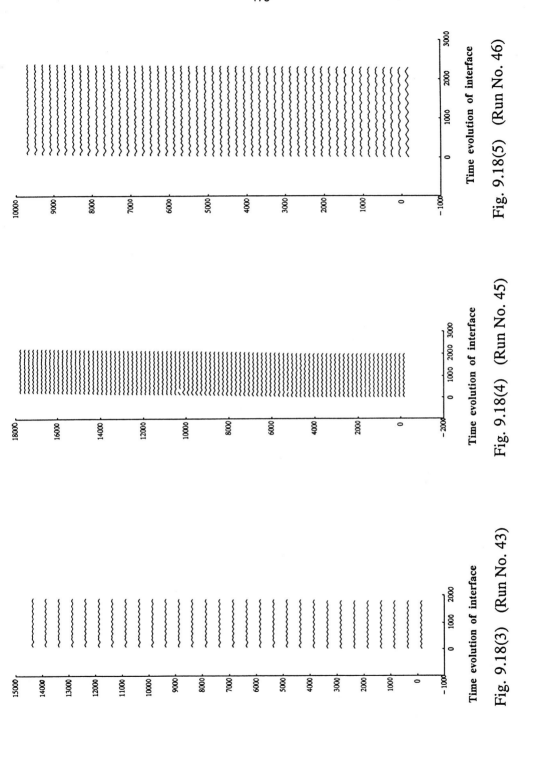

Fig. 9.18(3) (Run No. 43)

Fig. 9.18(4) (Run No. 45)

Fig. 9.18(5) (Run No. 46)

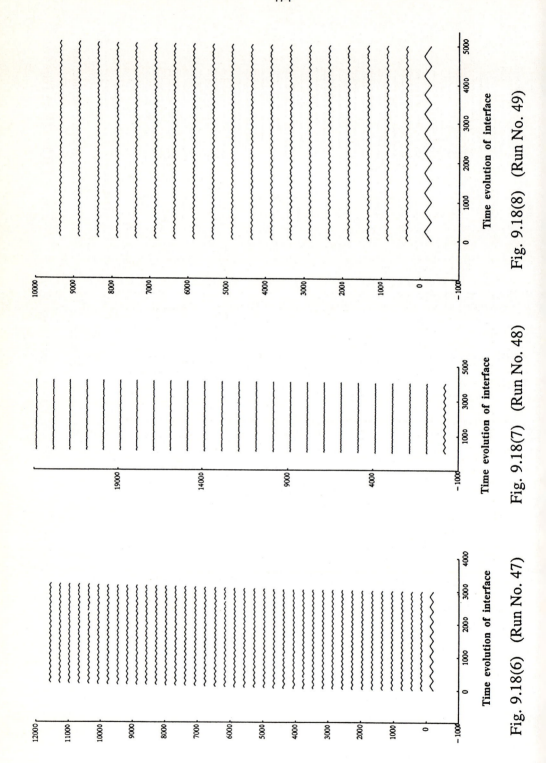

Fig. 9.18(6) (Run No. 47)   Fig. 9.18(7) (Run No. 48)   Fig. 9.18(8) (Run No. 49)

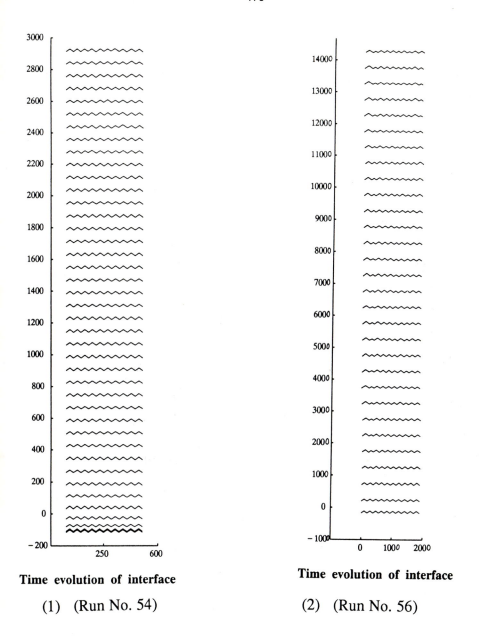

Fig. 9.19 Time evolution of the interface (Numerical experiment No. 5). (μm)

Fig. 9.20 Time evolution of the interface (Numerical experiment No. 6).   (μm)

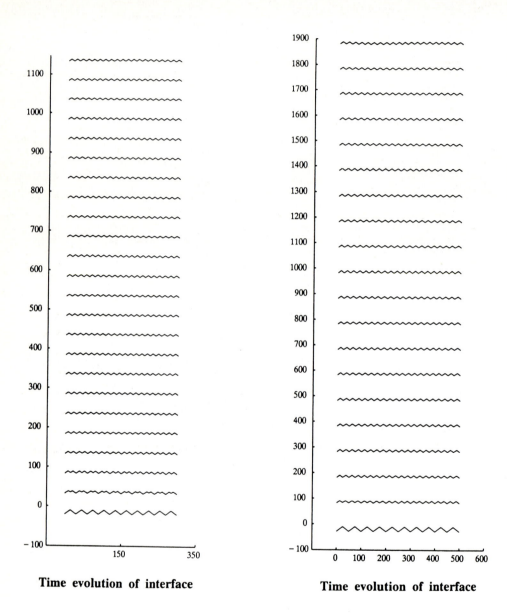

Time evolution of interface

Time evolution of interface

Fig. 9.20(1)   (Run No. 62)

Fig. 9.20(2)   (Run No. 63)

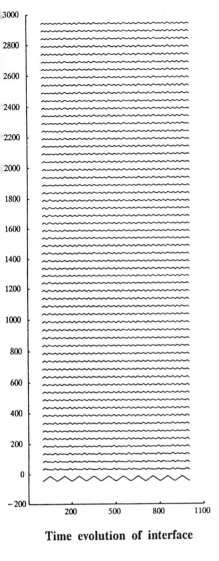

Time evolution of interface

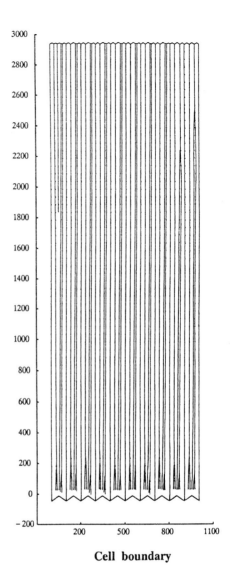

Cell boundary

Fig. 9.20(3) (Run No. 64) (a)

Fig. 9.20(3) (Run No. 64) (b)

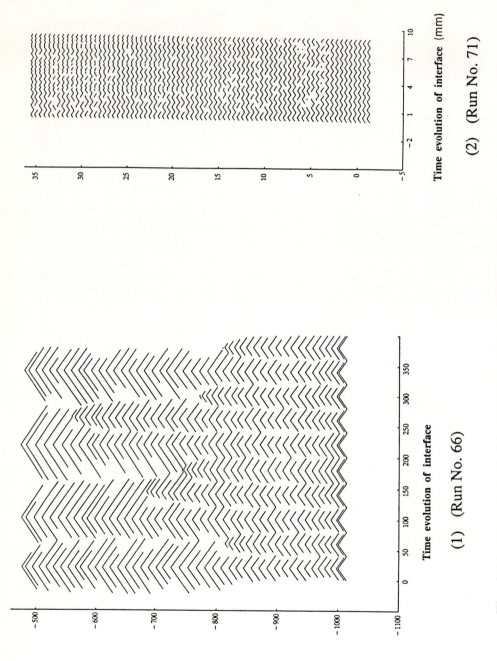

Fig. 9.21 Time evolution of the interface (Numerical experiment No. 7).

(1) (Run No. 66)

(2) (Run No. 71) (μm)

Fig. 9.22 Time evolution of the interface (Numerical experiment No. 8). (μm)

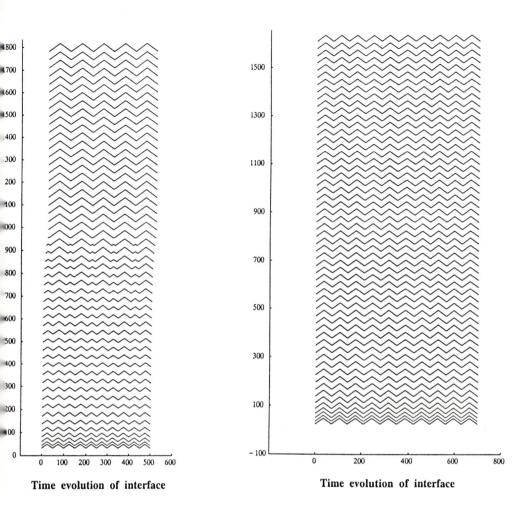

Fig. 9.22(1)　(Run No. 73)　　　Fig. 9.22(2)　(Run No. 74)

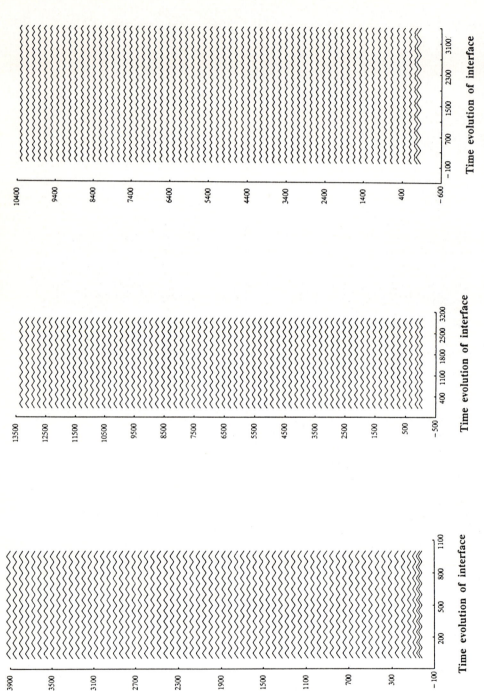

Fig. 9.22(3) (Run No. 75) Fig. 9.22(4) (Run No. 77) Fig. 9.22(5) (Run No. 78)

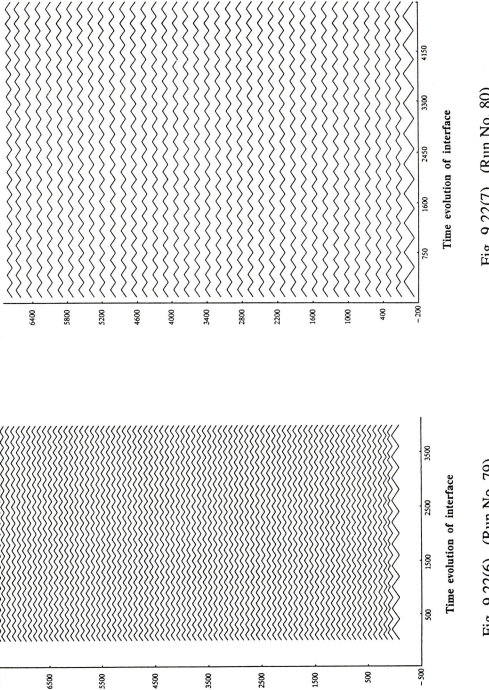

Fig. 9.22(6) (Run No. 79)

Fig. 9.22(7) (Run No. 80)

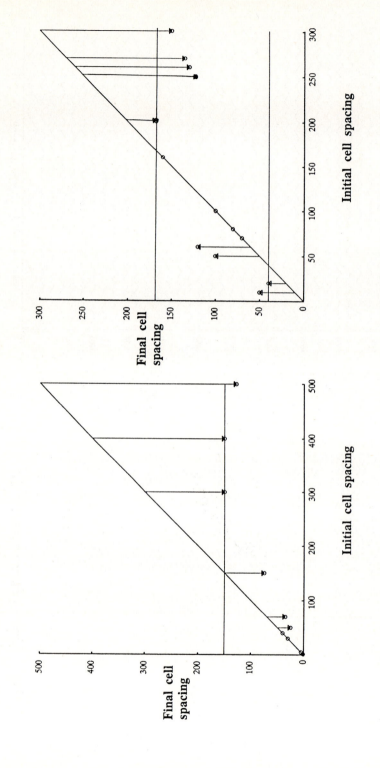

Fig. 9.23 The initial and final cell spacings. (μm)

Fig. 9.23(1) (Experiment No. 1)

Fig. 9.23(2) (Experiment No. 2)

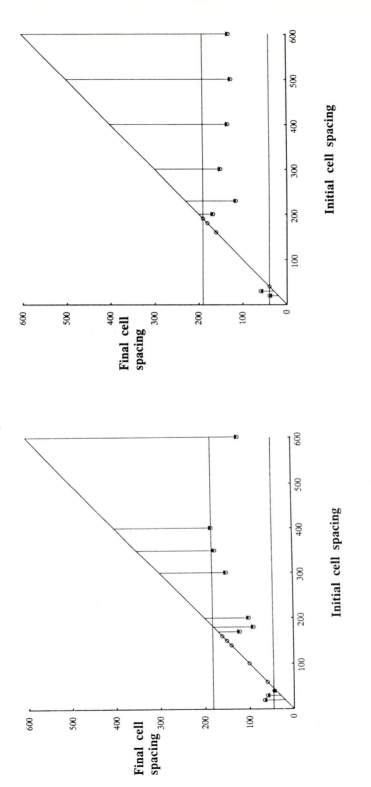

Fig. 9.23(3) (Experiment No. 3)

Fig. 9.23(4) (Experiment No. 4)

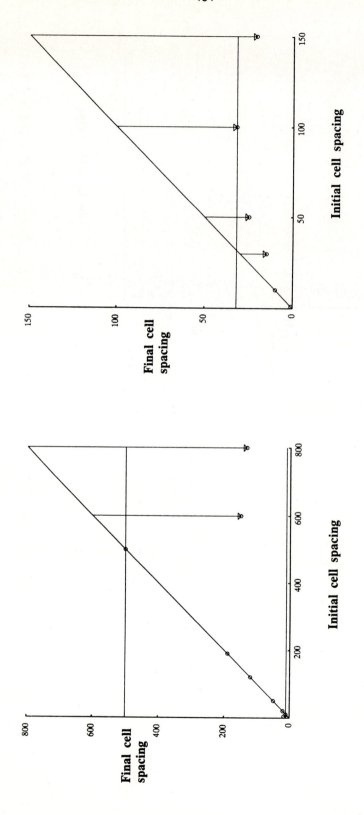

Fig. 9.23(5) (Experiment No. 5)

Fig. 9.23(6) (Experiment No. 6)

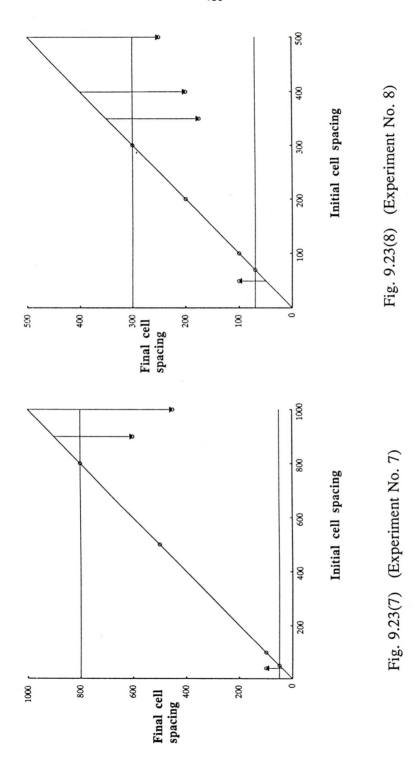

Fig. 9.23(8) (Experiment No. 8)

Fig. 9.23(7) (Experiment No. 7)

9.3.2 Cellular interactions

The results show that there are two types of cellular interactions: transient interactions and persistent interactions. As can be seen, in most of the runs, transient cellular interactions occur at the beginning of the growth process, after which the cells remain stable. These transient interactions occur either because the initial cell spacing is outside the stable spacing range, or because the initial condition is not the steady state condition. For example, in some cases, although the initial cell spacing is within the stable spacing range, cell splitting may occur during the transient period, but the cells created through the cell tip splitting cannot survive the steady state, changing the cell spacing back to the original spacing (*e.g.*, Figs. 9.16(3), 9.16(4) & 9.17(4)). It is also possible that an initial cell spacing, which is within the stable spacing range under that growth condition, may end up with another stable spacing within the stable cell spacing range (*e.g.*, Figs. 9.15(2), 9.16(2), 9.17(5) & 9.20(1)). Sometimes the cell tip splitting overshoots, and some of the cells created through tip splittings are lost later (*e.g.*, Figs. 9.15(4), 9.17(6) & 9.20(3)). The transient cellular interactions result in an upper limit and a lower limit to the cell size.

Under some conditions, however, the cellular interactions become persistent throughout the growth process (Fig. 9.15(3)). This is very interesting as it represents the strong cellular interactions observed in silicon under certain growth conditions (see Chapter 2).

It has not yet been established under what conditions this type of persistent interactions should occur; however, careful examinations on this growth process suggest the following features: (1) The growth condition is close to the critical constitutional undercooling condition; (2) The cells present are shallow cells. This is not surprising since intuitively it may be postulated that the deep liquid grooves along the cell boundary may stabilize the cells, because neighbouring cells will affect one another to a much less extent as the liquid groove plays a more dominant role in solute flow making the solute flow sideways less important; (3) The maximum kinetic undercooling along the facet has been observed to be somewhere midway along the facet (Fig. 9.24), perhaps closer to the base of the cell than the tip. This means that the sites of the maximum kinetic undercooling along two neighbouring facets are on the one hand away from each other, such that their growth rates, determined by the maximum kinetic undercooling, can differ from each other (which is the source for their growth competition), on the other hand they are still close enough for the two facets to influence each other; (4) It appears that the cellular interactions occur periodically during the growth process, *i.e.*, the interaction persists for approximately $D/(kV)$, which is the transient diffusion distance, and then the interface remains stable for about D/V, which is the thickness of the diffusion boundary layer, before a new period of interaction starts. The underlying mechanism is yet to be understood. It is hoped that through more numerical experiments it will be possible to establish clearly how the cellular interactions vary with the growth condition.

9.3.3 The stable cell spacing range

In Fig. 9.25 the stable cell spacing range is summarized for different growth conditions. It can be seen that the range of stable cell spacings are wider at very high or very low growth rates and narrower at intermediate growth rates (Fig. 9.25(a)).

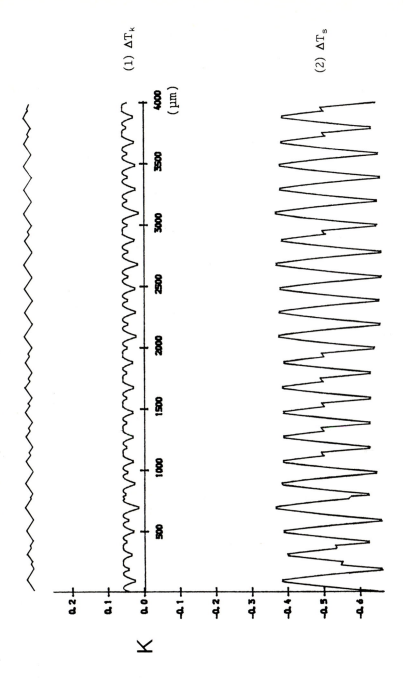

Fig. 9.24 The interface undercooling at the end of Run 9 (as shown in Fig. 9.15(3)).

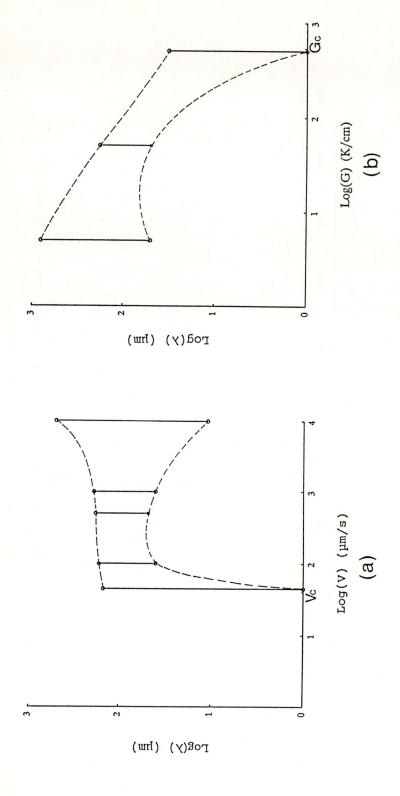

Fig. 9.25 The stable cell spacing range. (a) G = 50 K/cm (b) V = 500 μm/s

Under the same temperature gradient, the effect of the growth velocity on the cell spacing is two-folded: on the one hand, a larger velocity requires a larger kinetic undercooling to maintain the steady state growth rate thus requiring a larger facet; on the other hand, a larger velocity requires smaller cell spacings to ease the diffusional requirements. The range of stable cell spacings is the result of the interplay between these two factors.

It also appears that the range of stable cell spacings is narrower at the medium temperature gradient and wider at the two extremes (Fig. 9.25(b)).

9.3.4 Solute redistribution

Fig. 9.26 shows examples of the calculated solute field after the system reaches the steady state. The isoconcentrates are obtained by interpolating between grid-point compositions linearly along X and exponentially along Z.

The steady state solute profile is the result of the interplay between the cell spacing and D/V. As Fig. 9.26 shows, with a large Peclet number, $p = \lambda V/(2D)$, the cell spacing is large compared with the thickness of the solute boundary layer, and the isoconcentrates tend to be parallel to the interface (Fig. 9.26(1)). On the other hand, with a small Peclet number, the isoconcentrates tend to follow those for a planar interface (Fig. 9.26(2)), and the solute enrichment at the cell boundary will be severer.

9.3.5 Deep cells and shallow cells

The calculated time evolution of the interface under different growth conditions shows that, for $k > 1$, liquid grooves do not form and shallow cells are retained. This can also be seen from the steady state interface undercooling (Fig. 9.27). For $k < 1$, however, shallow cells are retained only at very low or very high growth rates; at intermediate growth rates liquid grooves tend to form (Fig. 9.28).

As has been mentioned earlier, liquid grooves form at cell boundaries because of severe solute enrichment. Figs. 9.29 - 9.31 show the steady state interface undercoolings calculated at three different growth rates. At a very low growth rate, solute enrichment is of less importance, and the kinetic undercooling at the cell base will remain above 0, therefore the interface will not break down (Fig. 9.29). At a very high growth rate, the kinetic undercooling required to maintain the steady state growth rate is relatively high (as can be seen from Eq. (9.1)), such that the kinetic undercooling at the cell base remains above 0, therefore shallow cells are again retained (Fig. 9.30). At an intermediate growth rate, however, the solute enrichment at the cell boundary reduces the kinetic undercooling significantly, such that the interface breaks down forming liquid grooves along the cell boundary (Fig. 9.31).

9.3.6 The effect of solute on cellular interactions

The results show that the solute effect plays an important role in the pattern formation of cellular array growth. Without solute effect shallow cells should always be retained, and the maximum kinetic undercooling should always occur at the base of the facet, so that neighbouring facets should have the same growth rates, and no interactions should occur.

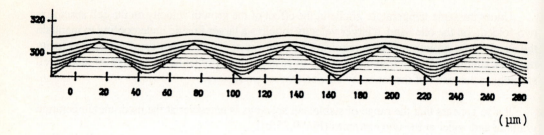

Fig. 9.26(1)　Isoconcentrates (p = 4.3) (Run 25).

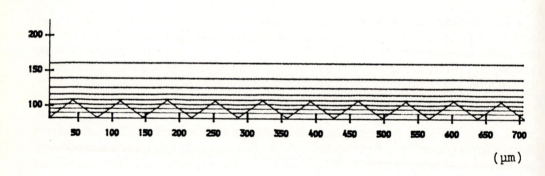

Fig. 9.26(2)　Isoconcentrates (p=1) (Run 15).

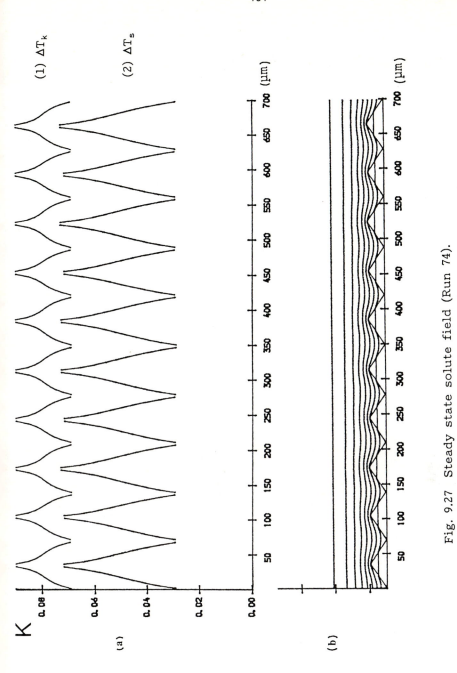

Fig. 9.27 Steady state solute field (Run 74).

(a) Interface undercooling.

(b) Isoconcentrates.

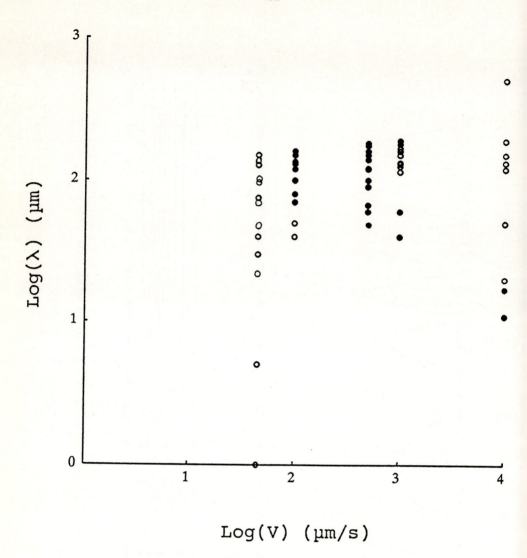

Fig. 9.28 Deep cells and shallow cells formed at different growth rates (G = 50 K/cm).
• Deep cells   ○ Shallow cells

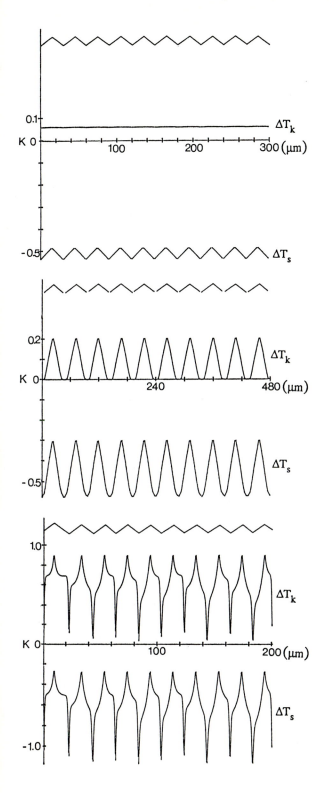

Fig. 9.29 Calculated steady state interface undercooling (k = 0.2, C_0 = 0.02%, V = 45 μm/s, G = 50 K/cm).

Fig. 9.30 Calculated steady state interface undercooling (k = 0.2, C_0 = 0.02%, V = 500 μm/s, G = 50 K/cm).

Fig. 9.31 Calculated steady state interface undercooling (k = 0.2, C_0 = 0.02%, V = 10 mm/s, G = 50 K/cm).

When solute is present, the maximum kinetic undercooling for each facet is normally not at the base of the facet, and neighbouring facets will have different growth rates. Different facet growth rates will lead to cellular interactions in the array. For example, if somehow a cell lags behind its neighbours (*e.g.* the third cell from the left in the array indicated by the arrow in Fig. 9.32), solute rejected from the neighbour cells will be dumped into the liquid region ahead of this cell, as indicated by the calculated isoconcentrates and the interface undercooling terms, ΔT_k and ΔT_s. A smaller ΔT_k (due to a larger ΔT_s) makes this cell grow even more slowly and lag further behind; eventually it is grown out.

9.3.7 Comparison with experimental observations

Experimental observations in thin film silicon [93-96] (see Chapter 2) suggest that the cellular interactions are stronger under some growth conditions resulting in more branched network of cell boundaries, but sometimes less dramatic resulting in non-branched, parallel cell boundaries (Fig. 9.33). This is in agreement with the results of this model, which predict two types of cellular interactions, *i.e.* transient interactions and persistent interactions, leading to the formations of two types of cell boundaries. Most of the experimental observations with faceting organic compounds described previously (see Chapter 5) in fact represent the less dramatic cellular interactions. The numerical results also explain why a unique stable cell spacing cannot be found in the experimental work (see Chapter 5) under a given growth condition. Solute enrichment along cell boundaries as is shown by the numerical results ($k < 1$) also agrees with the experimental work (see Chapter 5) and observations in silicon [97-98].

9.4 General Discussions

In Table 9.5 a comparison is made between the four numerical models. As can be seen none of the first three models are satisfactory. Model I produces persistent cellular interactions at all times resulting in a unique steady state average cell spacing. Models II & III produce stable structures at all times and predict an upper limit to the cell size. The very success of Model IV lies in the fact that it predicts the variation of cellular interactions with the growth condition, and an upper limit and a lower limit to the cell size. This is exactly what happens in practice. The success of the model is due to the fact that it includes the solute effect and the more realistic screw dislocation growth mechanism. It becomes apparent at this stage why it is so crucial to include the effect of solute.

It is noted that there are three variables in the growth condition, *i.e.*, V, G, and C_0, each of which may vary independently over several orders of magnitude. The establishment of a full picture of faceted cellular array growth must therefore await more thorough numerical experiments. It should also be interesting to carry out numerical experiments with a much higher external noise level. As an example, Fig. 9.34 shows the calculated growth process with tip perturbations applied periodically during the growth process. Perhaps this can test the stability of the interface more severely.

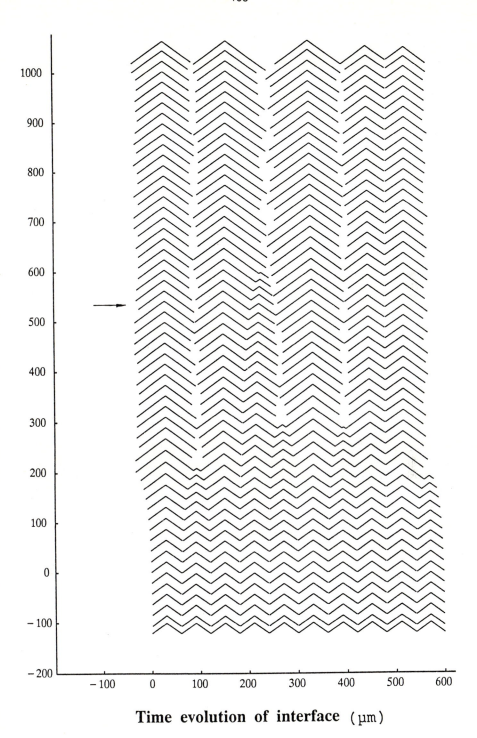

Time evolution of interface (μm)

Fig. 9.32(1) (Run 14)

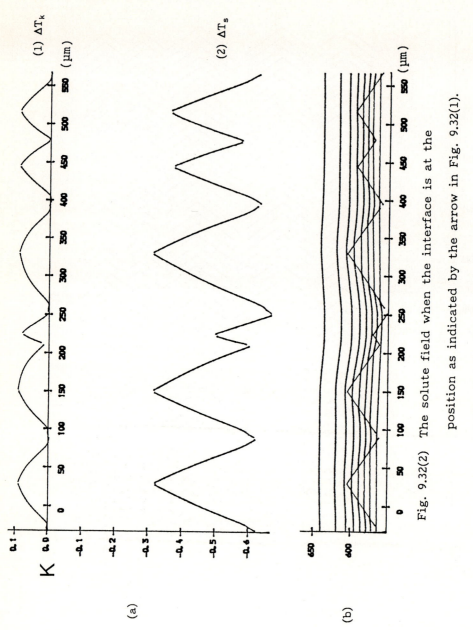

Fig. 9.32(2) The solute field when the interface is at the position as indicated by the arrow in Fig. 9.32(1).

(a) Interface undercooling.
(b) Isoconcentrates.

Table 9.5 Comparison of numerical models

Model	Growth kinetics	Solute effect	Cellular interface	Cellular interactions	Cell spacings
I	2-D nucleation / linear	No	Shallow	Persistent	Unique average spacing
II	2-D nucleation / exponential	No	Shallow	Transient	$\lambda < \lambda_{max}$
III	Growth by screw dislocations	No	Shallow	Transient	$\lambda < \lambda_{max}$
IV	Growth by screw dislocations	Solute (a) $k > 1$ (b) $k < 1$	Shallow Deep + shallow	Transient + persistent	$\lambda_{min} < \lambda < \lambda_{max}$

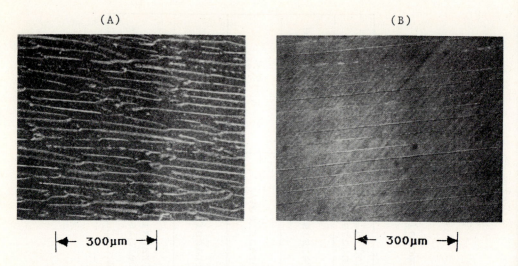

Fig. 9.33 Two types of cell boundary networks observed in thin film Si (taken from Ref. 94). (A) Branched (B) Non-branched

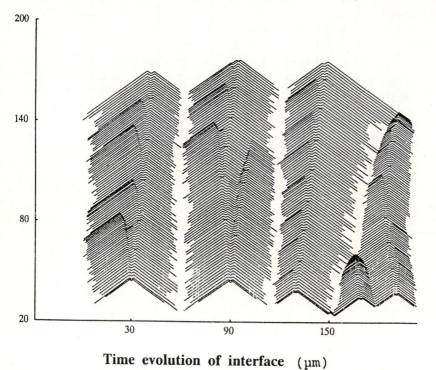

Time evolution of interface (μm)

Fig. 9.34 Time evolution of the interface when tip perturbations are applied periodically during the growth process.

9.5 Pattern Formation in Perspective

The present model has demonstrated that there are cellular interactions in the array which determine the pattern formation. As a result, there is a finite range of stable cell spacings under a given growth condition. This is different from previous theories, *e.g.*, the extremum growth hypothesis or the marginal stability criterion, which predict a unique steady state cell spacing (see Chapter 2).

Cells in an array interact because their solute field overlaps. Depending on the current state of the interface as compared with the steady state, any infinitesimal perturbation in the array can either be amplified while being transmitted laterally through the array leading to the interaction between cells, or die out during the growth process so that the interface remains stable. The results have shown that the cellular interaction in the array is stronger under some growth conditions, and less dramatic under other growth conditions. It has not yet been established precisely how this varies with the growth condition.

A generally similar behaviour of pattern formation can be expected for non-faceted cellular growth and eutectic growth. However it is noted that faceted growth differs from non-faceted growth in that the growth kinetics play a more dominant role. The relatively large kinetic undercooling term will not only make the planar interface more stable, but also makes it more difficult for faceted cells to adjust their growth condition. Therefore a wider range of stable cell spacings should be expected for faceted growth than for non-faceted growth.

It can be seen that it is essential to consider the array in order to describe cellular growth correctly and to understand the pattern formation of cellular growth. It can also be seen that the pattern formation problem should be approached in a vigorously dynamical fashion.

9.6 Conclusion

Numerical models have been developed to follow the time evolution of faceted cellular array growth. For the first time, the true time-dependent faceted cellular array growth has been modelled properly. Numerical experiments have been performed under different growth conditions. Numerical dynamical study of faceted cellular array growth has shown that the pattern formation is determined by cellular interactions in the array; there is a finite range of stable cell spacings under a given growth condition; and the range of stable cell spacings varies with the growth condition. It is suggested that the problem of pattern formation of cellular array growth should be approached in a vigorously dynamical fashion.

Chapter X

SUMMARY OF CONCLUSIONS AND SUGGESTIONS FOR FUTURE WORK

10.1 Conclusions

(1) The direct observation of faceted cellular growth has revealed cellular interactions in the array. Tip splitting and loss of cells have been observed to be the two main mechanisms for the variation of cell spacings during growth. It has been found that there is no unique steady state cell spacing under the same growth condition.

(2) It can be concluded from the numerical dynamical study of faceted cellular array growth that the pattern formation is determined by cellular interactions in the array; there is a finite range of stable cell spacings under a given growth condition. It is suggested that a generally similar pattern formation behaviour should be expected for other array growth processes, such as non-faceted cellular or eutectic growth.

(3) Two types of cellular interactions have been identified: transient interactions, which lead to the formation of regular structures, and persistent interactions, which lead to the formation of irregular structures. These can be related to the two types of cell boundary networks, *i.e.* branched and parallel cell boundary networks, found in thin film silicon single crystals.

(4) The cases for $k > 1$ and $k < 1$ are not symmetric as far as solute flow is concerned. For $k > 1$, shallow cells are usually retained, whereas for $k < 1$, the tendency for liquid groove formation varies with the growth condition. The pattern formation behaviours of deep cells and shallow cells however are similar.

(5) Numerical study of heat flow in zone melting has shown that the moving boundary problem of zone melting can be solved using an implicit enthalpy method for a pure material with the latent heat incorporated.

(6) Theoretical study of solute flow in steady state cellular array growth has demonstrated that the point source technique is effective and may be more accurate and potentially more rapid than numerical techniques used in previous works.

(7) Preliminary work on the measurement of steady state non-faceted array cell shapes has been carried out.

10.2 Suggestions for Future Work

(1) Extensive, systematic experimental study of faceted cellular growth should be carried out over a wide range of growth conditions for different faceting materials to establish the pattern formation behaviour.

(2) More numerical experiments should be conducted over as wide a range of growth conditions as possible to investigate the pattern formation behaviour, especially how the range of stable cell spacings varies with the growth condition. Comparisons should be made with experimental results once available.

(3) The findings on the pattern formation behaviour of faceted cellular growth, once established, should be applied to thin film silicon single crystals aiming at producing large area defect-free crystals.

(4) A similar approach can be applied to non-faceted cellular array growth and eutectic growth. More careful experimental work with non-faceted cellular growth is needed in order to establish the true picture of pattern formation of non-faceted cellular growth.

(5) The point source technique should be further developed such that the interface undercooling condition can be incorporated, and transient problems can be dealt with. This is potentially a very powerful technique for moving boundary problems.

(6) The preliminary experimental work with steady state non-faceted cellular growth should be further developed.

More recent developments in the field of pattern formation in cellular growth can be found in references [120-127] and others.

REFERENCES

1. Jackson, K. A.: Liquid Metals and Solidification, p. 174, American Society for Metals, Cleveland, Ohio (1958).

2. Jackson, K. A.: Crystal Growth, Proc. Int. Conf. on Crystal Growth, Boston, 20-24 June, 1966. (Supplement to Physics & Chemistry of Solids), p. 17.

3. Woodruff, D. P.: The Solid-Liquid Interface, p. 46, Cambridge University Press (1973).

4. Jackson, K. A. & Hunt, J. D.: Acta Metall., 13, 1212 (1965).

5. Cahn, J. W.: Acta Metall., 8, 554 (1960).

6. Jackson, K. A., Uhlmann, D. R. & Hunt, J. D.: J. Crystal Growth, 1, 1 (1967).

7. Christian, J. W.: The Theory of Transformations in Metals and Alloys, Pergamon Press (1965).

8. Flemings, M. C.: Solidification Processing, p. 290, McGraw-Hill, New York (1974).

9. Frank, F. C.: Discussions Faraday Soc., 5, 48 (1949).

10. Burton, W. K., Cabrera, N. & Frank, F. C.: Phil. Trans., A243, 299(1950).

11. Gilmer, G. H. & Bennema, P.: J. Crystal Growth, 13/14, 148(1972).

12. Jackson, K. A.: J. Crystal Growth, 24/25, 11(1974).

13. Weeks, J. D. & Gilmer, G. H.: Adv. Chem. Phys., 40, 157 (1979).

14. Jackson, K. A.: Materials Science and Engineering, 65(1), 7(1984).

15. Cahn, J. W., Hillig, W. B. & Sears, G. W.: Acta Metall., 12, 1421 (1964).

16. Abbaschian, G. J. & Ravitz, S. F.: J. Crystal Growth, 44, 453(1978).

17. Peteves, S. D., Alvarez, J. & Abbaschian, G. J.: Rapidly Solidified Metastable Materials (Eds.: Kear, B.H. & Geissen, B. C.), p. 15, Elsevier, New York (1984).

18. Peteves, S. D. & Abbaschian, G. J.: Interfacial Kinetics and Roughening, Fall' 85 AIME-TMS Meeting, Toronto, Canada.

19. Peteves, S. D. & Abbaschian, G. J.: J. Crystal Growth, 79, 775 (1986).

20. Jackson, K. A. & Hunt, J. D.: Trans. Met. Soc. AIME, 236, 246 (1966).

21. O'Hara, S., Tarshis, L. A., Tiller, W. A. & Hunt, J. D.: J. Crystal Growth, 3/4, 555(1968).

22. Herring, C.: Chapter in Structure and Properties of Solid Surfaces (Eds.: Gomer, R. & Smith, C. S.), Univ. of Chicago Press, Chicago (1952), p. 68.

23. Mullins, W. W. & Sekerka, R. F.: J. Appl. Phys., 35, 444 (1964).

24. Gionanola, B., Kurz, W. & Trivedi, R.: Solidification Processing 1987 (Eds. Beech, J. and Jones, H.), The Institute of Metals, London (1988), p.246.

25. Gilmer, G. H.: Materials Science and Engineering, 65(1), 15(1984).

26. Boettinger, W. J. & Coriell, S. R.: Materials Science and Engineering, 65(1), 27(1984).

27. Jones, H.: Materials Science and Engineering, 65(1), 145(1984).

28. Poate, J. M.: J. Crystal Growth, 79, 549(1986).

29. Favier, J. J., Hunt, J. D. & Sahm, P. R.: Fluid Sciences and Materials Science in Space, p. 477, Springer-Verlag (1987).

30. Carslaw, H. S. & Jaeger, J. C.: Conduction of Heat in Solids, Oxford University Press (1959).

31. Tiller, W. A., Jackson, K. A., Rutter, J. W. & Chalmers, B.: Acta Metall., 1, 428 (1953).

32. Favier, J. J. & Camel, D.: J. Crystal Growth, 79, 50(1986).

33. Coriell, S. R.: 8th Int. Conf. on Crystal Growth, York, U.K., July 1986.

34. Brown, R. A.: Materials Sciences in Space, (Ed. Feuerbacher, B.), p. 55, Springer-Verlag, Berlin (1986).

35. Ungar, L. H. & Brown, R. A.: Phys. Rev., B29, 1367 (1984).

36. Ungar, L. H. & Brown, R. A.: Phys. Rev., B30, 3993 (1984).

37. Ungar, L. H. & Brown, R. A.: Phys. Rev., B31, 5931 (1984).

38. Ungar, L. H., Bennet, M. J. & Brown, R. A.: Phys. Rev., B31, 5923 (1985).

39. McFadden, G. B. & Coriell, S. R.: Physica, 12D, 253 (1984).

40. Burden, M. H.: D. Phil. Thesis, Oxford Univ. (1973).

41. Burden, M. H. & Hunt, J. D.: J. Crystal Growth, 22, 109 (1974).

42. Hunt, J. D.: Solidification and Casting of Metals, p. 3, Metals Soc., London (1979).

43. Bower, T. F., Brody, H. D. & Flemings, M. C.: Trans. Met. Soc. AIME, 236, 624 (1966).

44. Laxmanan, V.: Acta Metall., 33, 1023, 1037 (1985).

45. Kurz, W. & Fisher, D. J.: Acta Metall., 29, 11 (1981).

46. Trivedi, R.: J. Crystal Growth, 49, 219 (1980).

47. McCartney, D. G. & Hunt, J. D.: Metall. Trans., 15A, 983 (1984).

48. Hunt, J. D. & McCartney, D. G.: Acta Metall., 35, 89 (1987).

49. Temkin, D. E.: Dokl. Akad. Nauk. SSSR, 132, 1307(1960).

50. Bolling, G. F. & Tiller, W. A.: J. Appl. Phys., 32, 2587(1961).

51. Trivedi, R.: Acta Metall., 18, 287(1970).

52. Jordan, R. M. & Hunt, J. D.: Metall. Trans., 3, 1385(1972).

53. Tassa, M. & Hunt, J. D.: J. Crystal Growth, 34, 38(1976).

54. Langer, J. S. & H. Muller-Krumbhaar: Acta Metall., 26, 1681 (1978).

55. Langer, J. S. & H. Muller-Krumbhaar: Acta Metall., 26, 1689 (1978).

56. Langer, J. S. & H. Muller-Krumbhaar: Acta Metall., 26, 1697 (1978).

57. Huang, S.- C. & Glicksman, M. E.: Acta Metall., 29, 701 (1981).

58. Lipton, J., Glicksman, M. E. & Kurz, W.: Mater. Sci. Eng., 65, 57(1984).

59. McCartney, D. G.: D. Phil. Thesis, Oxford Univ. (1981).

60. McCartney, D. G. & Hunt, J. D.: Acta Metall., 29, 1851 (1981).

61. Burden, M. H. & Hunt, J. D.: J. Crystal Growth, 22, 99(1974).

62. Eshelman, M. A., Seetharaman, V. & Trivedi, R.: Acta Metall., 36(4), 1165(1988).

63. Venugopalan, D. & Kirkaldy, J. S.: Acta Metall., 32, 893(1984).

64. Sharp, R. M. & Hellawell, A.: J. Crystal Growth, 6, 253(1970).

65. Rutter, J. W. & Chalmers, B.: Can. J. Phys., 31, 15(1953).

66. Tiller, W. A. & Rutter, J. W.: Can. J. Phys., 34, 96(1956).

67. Jin, I. & Purdy, G. R.: J. Crystal Growth, 23, 37(1974).

68. Cheveigne, S. de, Guthmann, C. & Lebrun, M. M.: J. Crystal Growth, 73, 242(1985).

69. Somboonsuk, K., Mason, J. T. & Trivedi, R.: Acta Metall., 15A, 967(1986).

70. Klaren, C., Verhoeven, J. D. & Trivedi, R.: Metall. Trans., 11A, 1853(1980).

71. Mason, J. T., Verhoeven, J. D. & Trivedi, R.: J. Crystal Growth, 59, 516(1982).

72. Mason, J. T., Verhoeven, J. D. & Trivedi, R.: Metall. Trans., 15A, 1665(1984).

73. Esaka, H. & Kurz, W.: J. Crystal Growth, 72, 578(1985).

74. Bechhoefer, J. & Libchaber, A.: Phys. Rev., A35, 1393(1986).

75. Vinals, J., Sekerka, R. F. & Debroy, P. P.: J. Crystal Growth, 89, 405(1988).

76. Ivantzov, G. P.: Dokl. Akad Nauk SSSR, 58, 567(1947).

77. Horvay, G. & Cahn, J. W.: Acta Metall., 9, 695(1961).

78. Glicksman, M. E. & Schaefer, R. J.: J. Crystal Growth, 1, 297(1967).

79. Glicksman, M. E. & Schaefer, R. J.: J. Crystal Growth, 2, 239(1968).

80. Nash, G. E. & Glicksman, M. E.: Acta Metall., 22, 1283(1974).

81. Glicksman, M. E.: Mat. Sci. Eng., 65(1), 45(1984).

82. Trivedi, R. & Somboonsuk, K.: Mat. Sci. Eng., 65(1), 65(1984).

83. Bardsley, W., Callan, J. M., Chedzey, H. A. & Hurle, D. T. J.: Solid-State Electronics, 3, 142(1961).

84. Bardsley, W., Boulton, J. S. & Hurle, D. T. J.: Solid-State Electronics, 5, 395 (1962).

85. Bardsley, W., Mullin, J. B. & Hurle, D. T. J.: in 'The Solidification of Metals', p. 93, Iron and Steel Inst. Publication No. 110, London (1968).

86. Bardsley, W., Cockayne, B., Green, G. W. & Hurle, D. T. J.: Solid-State Electronics, 6, 389(1963).

87. Cockayne, B.: J. Crystal Growth, 4, 1(1968).

88. Tarshis, L. A. & Tiller, W. A.: Crystal Growth, Proc. Int. Conf. Cryst. Growth, Boston, 20-24 June, 1966, (Supplement to Physics and Chemistry of Solids), p. 709.

89. Chernov, A. A.: J. Crystal Growth, 24/25, 11(1974).

90. Pfeiffer, L., Paines, S., Gilmer, G. H., Saarloos, W. & West, K. W.: Phys. Rev. Lett., 54(17), 1944 (1985).

91. Geis, M. W., Smith, H. I., Tsaur, B-Y., Fan, J. C. C., Silversmith, D. J. & Mountain, R. W.: J. Electrochem. Soc., 129(12), 2812(1982).

92. Im, J. S., Chen, C. K., Thompson, C. V., Geis, M. W. & Tomita, H.: Mat. Res. Soc. Symp. Proc., vol. 107, Silicon-on-Insulator and Buried Metals in Semiconductors (Eds.: Sturm, J. C., Chen, C. K., Pfeiffer, L. & Hemment, P. L. F.), p. 169.

93. Pfeiffer, L., Gelman, A. E., Jackson, K. A. & West, K. W.: Mat. Res. Soc. Symp. Proc., 74, 543 (1986) (Eds.: Picraux, S. T., Thomson, M. O. & Williams, J. S.).

94. Zavracky, P. M., Vu, D. P., Allen, L., Henderson, W., Guckel, H., Sniegowski, J. J., Ford, T. P. & Fan, J. C. C.: Mat. Res. Soc. Symp. Proc., vol. 107, Silicon-on-Insulator and Buried Metals in Semiconductors (Eds.: Sturm, J. C., Chen, C. K., Pfeiffer, L. & Hemment, P. L. F.), p. 213.

95. Dutartre, D.: Mat. Res. Soc. Symp. Proc., vol. 107, Silicon-on-Insulator and Buried Metals in Semiconductors (Eds.: Sturm, J. C., Chen, C. K., Pfeiffer, L. & Hemment, P. L. F.), p. 157.

96. Geis, M. W., Smith, H. I., Silversmith, D. J. & Mountain, R. W.: J. Electrochem. Soc., 130(5), 1178(1983).

97. Fan, J. C. C., Tsaur, B. - Y., Chen, C. K., Dick, J. R. & Kazmerski, L.L.: Mat. Res. Soc. Symp. Proc., 23, 477 (1984).

98. Mertens, P. W., Wouters, D. J. , Maes, H. E., Veirman, A. D. & Landuyt, J. V.: J. Appl. Phys., 63(8), 2660(1988).

99. Harada, H., Itoh, T., Ozawa, N. & Abe, T.: Proc. 3rd Int. Symp. VLSI Sci. Tech., VLSI Science and Technology, p. 526 (1985).

100. Baumgart, H.: Proc. SPIE. Int. Soc. Opt. Eng., Vol. 623, 211(1986).

101. Datye, V., Mathur, R. & Langer, J. S.: J. Statistical Physics, 29(1), 1(1982).

102. Karma, A.: Phys. Rev. Letts., 59(1), 71(1987).

103. Catalogue Handbook of Fine Chemicals, Aldrich Chemical Company Limited, 1988.

104. Gutzon, I.: J. Crystal Growth, 42, 15(1977).

105. Scherer, G., Uhlmann, D. R., Miller, C. E. & Jackson, K. A.: J. Crystal Growth, 23, 323(1974).

106. Handbook of Chemistry and Physics, 47th Edition, CRC Press (1966).

107. Handbook of Chemistry and Physics, 68th Edition, CRC Press (1987).

108. Beisteins Handbuch der Organischen Chemie, 4th Edition.

109. Pfann, W. G.: Trans. AIME, 194, 747(1952).

110. Pfann, W. G.: Zone Melting (2nd Edition), John Wiley and Sons, Inc., New York (1966).

111. Herington, E. F. G.: Zone Melting of Organic Compounds, John Wiley and Sons, Inc., New York (1963).

112. Kobayashi, N.: J. Crystal Growth, 43, 417(1978).

113. Kuiken, H. K. & Roksnoer, P. J.: J. Crystal Growth, 47, 29(1979).

114. Otani, S., Tanaka, T. & Ishizawa, Y.: J. Crystal Growth, 66, 419(1984).

115. Otani, S., Tanaka, T. & Ishizawa, Y.: J. Less-Common Metals, 113, 205(1985).

116. Braun, J. H. & Pellin, R. A.: J. Electrochem. Soc., 108(10), 969(1961).

117. Kreyszig, E.: Advanced Engineering Mathematics (5th Edition), John Wiley & Sons, Inc., (1983).

118. Abramowitz, M. & Stegun, I. A.: Handbook of Mathematical Functions, Dover Publications, Inc., New York (1972).

119. The NAG Fortran Library Manual, Mark 12, The Numerical Algorithms Group Limited, Oxford (1987).

120. Shangguan, D.K. & Hunt, J.D.: Paper No. 37, Metals and Materials '87, The Institute of Metals, April 1987, Southampton, U.K.

121. Shangguan, D.K. & Hunt, J.D.: Solidification Processing 1987 (Eds. Beech, J. and Jones, H.), The Institute of Metals, London (1988), pp. 206-209.

122. Shangguan, D.K. & Hunt, J.D.: Silicon-On-Insulator and Buried Metals in Semiconductors, Materials Research Society Symposium Proceedings, 1987, vol. 107 (Eds. Sturm, J.C., Chen, C.K., Pfeiffer, L., and Hemment, P.L.F.), pp. 175-181.

123. Shangguan, D.K. & Hunt, J.D.: J. Crystal Growth, 96, 856(1989).

124. Shangguan, D.K. & Hunt, J.D.: Metall. Trans., 22A, 941(1991).

125. Shangguan, D. & Hunt, J.D.: 1991 TMS Annual Meeting, February 1991, New Orleans, Louisiana. To appear in Metall. Trans.

126. Rauh, H. & Hunt, J.D.: Solidification Processing 1987 (Eds. Beech, J. and Jones, H.), The Institute of Metals, London (1988), pp. 366-369.

127. Shangguan, D. & Hunt, J.D.: Metall. Trans., 22A, 1683(1991).

Subject Index

Atomic process 2
Array growth 17, 92
  non-faceted cellular 17, 70
  faceted cellular 19, 49, 123, 140, 156
Boundary condition 80, 94, 140
Cahn's theory 4
Cellular interaction 57, 186, 189, 200
Consistency 26
Constitutional undercooling 16
Continuous growth 6
Convergency 26
Corrector-predictor method 86
Crank-Nicholson scheme 26, 27
Deep cell 52, 189, 200
Differential equation 22, 76, 92, 94, 140
Diffusion equation 92, 94, 140
Discretisation equation 22, 78, 112, 146
Enthalpy method 83
Extrapolation 23
Extremum growth hypothesis 18
Faceted growth 6
Finite difference method 22
  explicit 23, 141
Finite element method 22
Fourier transform 96
Gaussian elimination 87, 113
Green's function 96
Green's theorem 100
Heat and solute transport 15
Heat flow 76
Implicit scheme 26, 27, 85
Integral equation 108

Interface kinetics 6
Interface structure 2
Interface undercooling 14, 121, 125
Interpolation 23, 154, 156
Isolated dendritic growth 17, 19
Jackson's theory 2
Kummer's acceleration method 106
Lateral growth 8
Liquid groove 52, 153, 189
Loss of cells 57, 128, 153, 200
Marginal stability 18
Molecular dynamics 11
Mullins-Sekerka theory 16
Newton-Raphson method 86
Non-faceted growth 6, 70
Numerical integration 113
Numerical modelling 22
Pattern formation 1, 21, 199, 200
Planar interface 1, 16, 49, 70, 121, 141
Point source technique 92
Screw dislocation 8, 139, 140
Shallow cell 52, 189, 200
Solute redistribution 52, 189
Stability 16, 17, 26
Stable cell spacing range 186
Steady state 23, 88, 92, 128
Temperature gradient stage 28
Tip splitting 57, 128, 154
Two dimensional nucleation 8, 135
Vacuum distillation 33
Zone melting 19, 76
Zone refining 37